Make an Arduino-Controlled Robot

Michael Margolis

O'REILLY®

Beijing · Cambridge · Farnham · Köln · Sebastopol · Tokyo

Make an Arduino-Controlled Robot

by Michael Margolis

Published by O'Reilly Media, Inc., 1005 Gravenstein Highway North, Sebastopol, CA 95472.

O'Reilly books may be purchased for educational, business, or sales promotional use. Online editions are also available for most titles (*http://my.safaribooksonline.com*). For more information, contact our corporate/institutional sales department: 800-998-9938 or *corporate@oreilly.com*.

Editor: Brian Jepson	**Production Editor:** Rachel Steely
	Interior Designers: Nellie McKesson and Edie Freedman

October 2012:   First Edition

Revision History for the First Edition:

2012-09-12   First release

2012-10-03   Second release

See *http://oreilly.com/catalog/errata.csp?isbn=9781449344375* for release details.

ISBN: 978-1-449-34437-5

[LSI]

Table of Contents

Preface

Building a robot and enabling it to sense its environment is a wonderful way to take your Arduino knowledge to the next level. In writing this book, I have brought together my love for invention and my experience with electronics, robotics and microcontrollers. I hope you have as much pleasure building and enhancing your robot as I did developing the techniques contained in this book.

Arduino is a family of microcontrollers (tiny computers) and a software creation environment that makes it easy for you to create programs (called *sketches*) that can interact with the physical world. Arduino enables your robot to sense the environment and respond in a rich variety of ways. This book helps you to build a robot that is capable of performing a wide variety of tasks. It explains how to assemble two of the most popular mobile platforms, a robot with two wheels and a caster (for stability, since it's hard to balance on two wheels), and a robot with four wheels and motors. If you want your robot up and running quickly, choosing one of the kits detailed in this book should speed you through the build process and get you going with the robot projects. But whether you prefer to design and build a platform of your own construction or build from a kit, you will find the projects that comprise the core of this book a practical and fun introduction to Arduino robots.

Who This Book Is For

This book is for people who want to explore robotics concepts like: movement, obstacle detection, handling sensors, remote control, and all kinds of real world physical computing challenges. It is for people who want to understand how these concepts can be used to build, expand and customize your robot. See "What Was Left Out" (page xi) for some general references for those with limited programming or electronics experience.

How This Book Is Organized

The book contains information that covers a broad range of robotics tasks. The hardware and software is built up stage by stage, with each chapter using concepts explained in earlier chapters. A simple "Hello Robot" sketch is introduced in Chapter 6, *Testing the Robot's Basic Functions* and extended in subsequent chapters. Each chapter introduces sketches that add new capabilities to the robot. Experienced users can skip directly to the chapters of interest—full source code for every sketch in this book is available online. However, users who want to learn all about the techniques covered will benefit and hopefully enjoy working with all the sketches presented in the book, as each sketch enables the robot to perform increasingly complex tasks.

The sketches are built using functional modules. The modules are stored using Arduino IDE tabs (see Chapter 5). Modules described in early chapters are reused later and to avoid printing the same code over and over in the book, only code that is new or changed is printed. Figure P-1 illustrates how the code is enhanced from sketch to sketch. The horizontal bars represent the sketches, the vertical bars represent functional modules that are included in the sketches. The initial 'helloRobot' sketch is transformed into the 'myRobot' sketch by the moving the code for program definitions into a module named *robotDefines.ino* and reflectance sensors into a module named *IrSensors.ino*. These module are included as tabs in the 'myRobot' sketch. Each subsequent sketch is enhanced by adding code to an existing module or creating a new module as a tab.

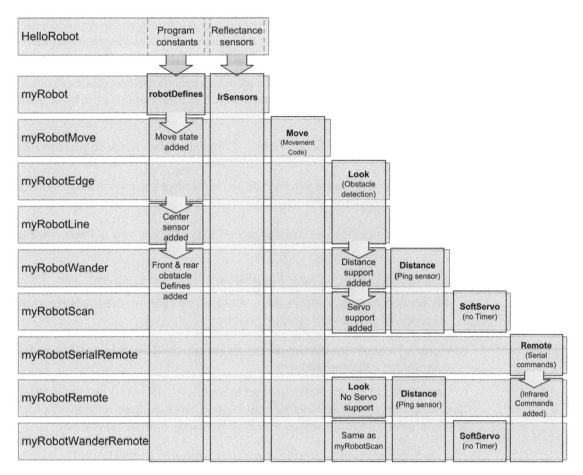

Figure P-1. *Sketch and module family tree*

All code for every sketch is available in the download for this book and you can load the sketch being discussed into your IDE if you want a complete view of all the code.

Chapter 1, *Introduction to Robot Building* provides a brief introduction to robot hardware and software.

Chapter 2, *Building the Electronics* describes how to prepare the electronics for use with the robot.

Chapter 3, *Building the Two-Wheeled Mobile Platform* describes how to assemble the 2 Wheel Drive (2WD) mobile platform.

Chapter 4, *Building the Four-Wheeled Mobile Platform* describes how to assemble the 4 Wheel Drive (4WD) mobile platform.

Chapter 5, *Tutorial: Getting Started with Arduino* introduces the Arduino environment and provides help getting the development environment and hardware installed and working.

Chapter 6, *Testing the Robot's Basic Functions* explains the first robotics sketch. It is used to test the robot. The code covered in this chapter is the basis of all other sketches in the book:

- *HelloRobot.ino* (Arduino sketch) — Brings the robot to life so you can test your build.
- *myRobot.ino* — Same functionality as above but structured into modules to make it easy to enhance.

Chapter 7, *Controlling Speed and Direction* explains how you make the robot move:

- *myRobotMove.ino* — Adds higher level movement capability.
- *myRobotCalibrateRotation.ino* — A sketch for running the robot through a range of speeds to calibrate the robot.

Chapter 8, *Tutorial: Introduction to Sensors* introduces the most popular sensors used with the 2WD and 4WD robots.

Chapter 9, *Modifying the Robot to React to Edges and Lines* describes techniques for using reflectance sensors to enable your robot to gain awareness of its environment. The robot will be able to follow lines or to avoid edges.

- *myRobotEdge.ino* — The robot will move about in an area bound by a non-reflective surface (a large sheet of white paper placed on a non-reflective surface).
- *myRobotLine.ino* — Repositions the sensors used above to allow the robot to follow black lines painted or taped to a white surface. A variant of this sketch that sends data over serial for display on an external serial device is named `myRobotLineDisplay` and is included in the download code.

Chapter 10, *Autonomous Movement* describes how to use distance sensors to enable the robot to see and avoid obstacles encountered as it moves around.

- *myRobotWander.ino* — Adds 'eyes' to give the robot the ability to look around and avoid obstacles.
- *myRobotScan.ino* — Adds a servo so robot 'eyes' can scan independent of robot movement.

Chapter 11, *Remote Control* describes techniques for remotely controlling the robot. Wired and wireless serial commands and using a TV type infrared remote control are covered.

- *myRobotSerialRemote.ino* — Controls the robot using serial commands.
- *myRobotRemote.ino* — Controls the robot using an IR remote controller.
- *LearningRemote.ino* — Captures key codes from your remote control to enable these to be added to the myRobotRemote sketch.
- *myRobotWanderRemote.ino* — Combines remote control with autonomous movement.

Appendix A, *Enhancing Your Robot* provides tips and techniques for designing and building complex projects.

Appendix B, *Using Other Hardware with Your Robot* describes some alternative solutions for motor control.

Appendix C, *Debugging Your Robot* has hardware and software debugging tips. This sections includes Arduino and Processing source code to enable real time graphical display of robot parameters on a computer screen.

- *myRobotDebug.ino* — Arduino example showing how to send data to your computer.
- *ArduinoDataDisplay.pde* (Processing sketch) — graphs data received from Arduino in real time.

Appendix D, *Power Sources* introduces some alternatives for powering your robot.

Appendix E, *Programming Constructs* provides a brief introduction to some of the programming constructs used in the sketches for this book that may not be familiar to some Arduino users.

Appendix F, *Arduino Pin and Timer Usage* summarizes the pins and Arduino resources used by the robot.

What Was Left Out

This book explains all the code used for the robot, but it is not an introduction to programming. If you want to learn more about programming with Arduino, you may want to refer to the Internet or to one of the following books:

- *Getting Started with Arduino, 2nd Edition* by Massimo Banzi (O'Reilly)
- *Arduino Cookbook, 2nd Edition* by Michael Margolis (O'Reilly)

A good book for inspiration on more robotics projects is:

- *Make: Arduino Bots and Gadgets* by Tero Karvinen, Kimmo Karvinen (O'Reilly)

Code Style (About the Code)

The code used throughout this book has been tailored to clearly illustrate the topic covered in each chapter. As a consequence, some common coding shortcuts have been avoided. Experienced C programmers often use rich but terse expressions that are efficient but can be a little difficult for beginners to read. For example, code that returns boolean values uses the somewhat verbose explicit expressions because they are easier for beginner programmers to read, see the example that follows, which returns true if no reflection was detected by the robot's sensor:

```
return irSensorDetect(sensor) == false;
```

Here is the terse version that returns the same thing (note the negation operator before the function call):

```
return !irSensorDetect(sensor);
```

Feel free to substitute your preferred style. Beginners should be reassured that there is no benefit in performance or code size in using the terse form.

One or two more advanced programming concepts have been used where this makes the code easier to enhance. For example, long lists of sequential constants use the enum declaration.

The enum keyword creates an enumeration; a list of constant integer values. All the enums in this book start from 0 and increase sequentially by one.

For example, the list of constants associated with movement directions could be expressed as:

```
const int MOV_LEFT = 0;
const int MOV_RIGHT = 1;
const int MOV_FORWARD = 2;
const int MOV_BACK = 3;
const int MOV_ROTATE = 4;
const int MOV_STOP = 5;
```

The following declares the same constants with the identical values:

```
enum {MOV_LEFT, MOV_RIGHT, MOV_FORWARD,
      MOV_BACK, MOV_ROTATE, MOV_STOP};
```

In addition to brevity, there are many advantages to the enum version of the code. If you want to know more about enum, an online search for c++ enum should tell you all you need to know and more.

Good programming practice involves ensuring that values used are valid (garbage in equals garbage out) by checking them before using them in calculations. However, to keep the code focused on the topic, error-checking code has been kept to a minimum. If you expand the code, you are encouraged to add error-checking where needed.

Arduino Hardware and Software

The examples in this book were built using the Arduino Leonardo and Uno boards (see Chapter 5). The code has been tested with Arduino release 1.0.1 (the first release that fully supports the Leonardo board). Although many of the sketches will run on earlier Arduino releases, this has not been tested. If you really want to use a release older than 1.0, you need to change the extension from *.ino* to *.pde* to load the sketch into a pre-1.0 IDE.

There is a website for this book where you can download code for this book; see "How to Contact Us" (page xv).

There is also a link to errata on that site. Errata give readers a way to let us know about typos, errors, and other problems with the book. Errata will be visible on the page immediately, and we'll confirm them after checking them out. O'Reilly can also fix errata in future printings of the book in electronic books, and on Safari® Books Online, making for a better reader experience pretty quickly.

If you have problems getting the code to work, check the web link to see if the code has been updated. The Arduino forum is a good place to post a question if you need more help: *http://www.arduino.cc*.

If you like—or don't like—this book, by all means, please let people know. Amazon reviews are one popular way to share your happiness or other comments. You can also leave reviews at the O'Reilly site for the book.

Conventions Used in This Book

The following font conventions are used in this book:

Italic

> Indicates pathnames, filenames, and program names; Internet addresses, such as domain names and URLs; and new items where they are defined

Constant width

Indicates command lines and options that should be typed verbatim; names and keywords in programs, including method names, variable names, and class names; and HTML element tags

Constant width bold

Indicates emphasis in program code lines

Constant width italic

Indicates text that should be replaced with user-supplied values

---

This icon signifies a tip, suggestion, or general note.

Using Code Examples

This book is here to help you make things with Arduino. In general, you may use the code in this book in your programs and documentation. You do not need to contact us for permission unless you're reproducing a significant portion of the code. For example, writing a program that uses several chunks of code from this book does not require permission. Selling or distributing a CD-ROM of examples from this book *does* require permission. Answering a question by citing this book and quoting example code does not require permission. Incorporating a significant amount of example code from this book into your product's documentation *does* require permission.

We appreciate, but do not require, attribution. An attribution usually includes the title, author, publisher, and ISBN. For example: "*Make an Arduino Controlled Robot* by Michael Margolis (O'Reilly). Copyright 2013 Michael Margolis, ISBN (978-1-4493-4437-5)."

If you feel your use of code examples falls outside fair use or the permission given here, feel free to contact us at *permissions@oreilly.com*.

Safari® Books Online

Safari Books Online is an on-demand digital library that lets you easily search over 7,500 technology and creative reference books and videos to find the answers you need quickly.

With a subscription, you can read any page and watch any video from our library online. Read books on your cell phone and mobile devices. Access new titles before they are available for print, and get exclusive access to manuscripts in development and post feedback for the authors. Copy and paste code samples, organize your favorites, download chapters, bookmark key sections, create notes, print out pages, and benefit from tons of other time-saving features.

O'Reilly Media has uploaded this book to the Safari Books Online service. To have full digital access to this book and others on similar topics from O'Reilly and other publishers, sign up for free at *http://my.safaribooksonline.com*.

How to Contact Us

We have tested and verified the information in this book to the best of our ability, but you may find that features have changed (or even that we have made a few mistakes!). Please let us know about any errors you find, as well as your suggestions for future editions, by writing to:

O'Reilly Media, Inc.
1005 Gravenstein Highway North
Sebastopol, CA 95472
800-998-9938 (in the United States or Canada)
707-829-0515 (international/local)
707-829-0104 (fax)

We have a web page for this book, where we list errata, example code, and any additional information. You can access this page at:

http://shop.oreilly.com/product/0636920028024.do

To comment or ask technical questions about this book, send email to:

bookquestions@oreilly.com

For more information about our books, courses, conferences, and news, see our website at *http://www.oreilly.com*.

Find us on Facebook: *http://facebook.com/oreilly*

Follow us on Twitter: *http://twitter.com/oreillymedia*

Watch us on YouTube: *http://www.youtube.com/oreillymedia*

Acknowledgments

Rob DeMartin, the business manager at Maker Media, was the driving force behind the botkits, which inspired the writing of this book. Isaac Alexander and Eric Weinhoffer at Maker Media ran with the concept to make it a product. I thank them for testing the content of the book to ensure that the projects and the hardware worked well together.

I am grateful to the Arduino community for contributing a wealth of free software, in particular, the IrRemote library from Ken Sherriff that is used in the remote control chapter. I would also like to express my appreciation to Limor Fried (Ladyada) for creating the hardware, software and online build notes for the motor shield used in this book.

Thanks also to DFRobot, the innovative company that designed the robot platforms and provided the exploded view drawings used in the build chapters.

Mat Fordy at Cool Components (coolcomponents.co.uk) organized the robotics workshop that provided a testing ground for the book's projects. It was helpful and rewarding to work with the participants, each with a different level of experience, to build the robots and see their pleasure in bringing their creations to life. Their feedback helped make the book content clear, practical and fun.

If I have achieved my goal of making the rich variety of technical topics in this book accessible to readers with limited electronics or programming experience, then much of the credit goes to Brian Jepson. Brian, who was also my editor for the Arduino Cookbook, was with me every step of the way. I thank him for his guidance: from his support and passion in beginning the project, to his editorial expertise and application of his masterful communications skills right through to using his technical knowledge to test all the projects in the book.

I would like to thank my entire family for listening to me explain the finer points of robotics during a week- long vacation in the early stages of preparing this book. Four generations of my family were patient and constructive at times when they would have preferred to be boating on the lake or walking in the woods.

Finally, this book would not be what it is without the contributions made by my wife, Barbara Faden. Her feedback on early drafts of the manuscript helped shape the content. I am especially grateful for her support and patience in the wake of disruption created as I wrangled with these two little robots to meet the book's deadline.

Introduction to Robot Building | 1

This book takes you through the steps needed to build a robot capable of autonomous movement and remote control. Build instructions are provided for 2WD (two wheel drive) and 4WD (four wheel drive) platforms. The platforms shown in Figure 1-1 and Figure 1-2 will make the construction a snap, but you can build your own robot chassis if you prefer. The connection and use of the control electronics and sensors are fully explained and the source code is included in the book and available for download online (see "How to Contact Us" (page xv) for more information on downloading the sample code).

Figure 1-1. *The assembled two wheeled robot chassis*

Figure 1-2. *The assembled four wheeled robot chassis*

Here is a preview of the projects you can build:

- Controlling speed and direction by adding high level movement capability.
- Enabling the robot to see the ground—using IR sensors for line and edge detection (see Figure 1-3 and Figure 1-4).
- Enabling the robot to look around—scanning using a servo so the robot can choose the best direction to move, as shown in Figure 1-5.
- Adding remote control using a TV remote control or a wired or wireless serial connection.

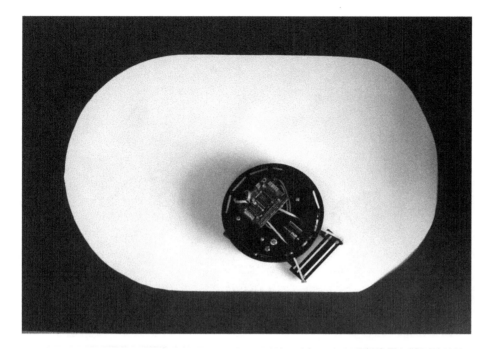

Figure 1-3. *Robot moves around but remains within the white area*

Figure 1-4. *Robot follows black line*

Figure 1-5. *Two wheeled and four wheeled robots with distance scanners*

Why Build a Robot?

Building a robot is different from any other project you can make with a microcontroller. A robot can move and respond to its environment and exhibit behaviors that mimic living creatures. Even though these behaviors may be simple, they convey a sense that your creation has a will and intent of its own. Building a machine that appears to have some spark of life has fascinated people throughout the ages. The robots built over 60 years ago by neurophysiologist W. Grey Walter (see *http://www.extremenxt.com/walter.htm*) explored ways that the rich connections between a small number of brain cells give rise to complex behaviors.

There are many different kinds of robots, some can crawl, or walk, or slither. The robots described in this book are the easiest and most popular; they use two or four wheels driven by motors.

Choosing Your Robot

The projects in this book can use either a two or four wheeled platform, but if you are still deciding which is right for you, here are some factors that will help you choose:

Two Wheeled Robot

Light and very maneuverable, this is a good choice if you want to experiment with tasks such as line-following that require dexterous movement. However, the caster that balances the robot requires a relatively smooth surface.

Four Wheeled Robot

This robot's four wheel drive makes this a good choice if you want it to roam over rougher surfaces. This platform has a large top plate that can be used to carry small objects. The robot is heavier and draws more current than the 2WD robot, so battery life is shorter.

How Robots Move

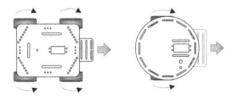

Figure 1-6. *Left and Right wheels turn forward, Robot moves Forward*

The robots covered in this book move forward, back, left and right much like a conventional car. Figure 1-6 shows the wheel motion to move the robot forward.

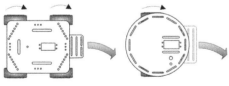

Figure 1-7. *Only Left wheels turn, Robot Turns Right*

If the wheels on one side are not driven (or are driven more slowly than the other side) the robot will turn, as in Figure 1-7.

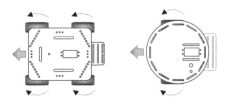

Figure 1-8 shows that reversing the wheel rotation drives the robot backward.

Figure 1-8. *Left and Right wheels turn backward, Robot moves Backward*

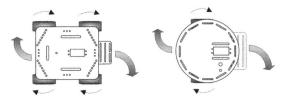

Unlike a car (but a little like a tank), these robots can also rotate in place by driving the wheels on each side in different directions. If the wheels on each side are spinning in opposite directions, the robot will rotate. Figure 1-9 shows clockwise rotation.

Figure 1-9. *Left wheels turn forward, Right wheels reverse, Robot rotates Clockwise*

Tools

These are the tools you need to assemble the robot chassis.

Phillips Screwdriver
   A small Phillips screwdriver from your local hardware store.

Small long-nose or needle-nose pliers
   For example, Radio Shack 4.5-inch mini long-nose pliers, part number 64-062 (see Figure 1-10) or Xcelite 4-inch mini long-nose pliers, model L4G.

Small wire cutters
   For example, Radio Shack 5" cutters, part number 64-064 (Figure 1-11) or Jameco 161411

Soldering iron
   For example, Radio Shack 640-2070 (Figure 1-12) or Jameco 2094143 are low cost irons suitable for beginners. But if you are serious about electronics, a good temperature controlled iron is worth the investment, such as Radio Shack 55027897 or Jameco 146595.

Solder 22 AWG (.6mm) or thinner
   For example, Radio Shack 640-0013 or Jameco 73605.

Figure 1-10. *Small Pliers*

Figure 1-11. *Wire Cutters (Side Cutters)*

Figure 1-12. *Soldering Iron*

Building the Electronics | 2

This chapter guides you through the electronic systems that will control your robot. Both the two wheeled and four wheeled platforms use the same modules, a pre-built Arduino board (Arduino Uno or Leonardo), and a motor controller kit. The motor controller featured in this book is the AFMotor shield from Adafruit Industries. Although other motor controllers can be used (see Appendix B) the AFMotor shield provides convenient connections for the signals and power to all the sensors and devices covered in this book. It is also capable of driving four motors, which is required for the four wheel drive chassis.

Although the attachment of the boards to the robot differs somewhat depending on the chassis, the building of the AFMotor circuit board kit is the same for both. If you don't have much experience with soldering, you should practice soldering on some wires before tackling the circuit board (you can find soldering tutorials here: *http://www.ladyada.net/learn/soldering/thm.html*).

Hardware Required

See *http://shop.oreilly.com/product/0636920028024.do* for a detailed parts list.

- Tools listed in "Tools" (page 6)
- AFMotor shield kit
- Three 6 way 0.1" female headers
- Three QTR-1A reflectance sensors
- Stripboard, three 3 way 0.1" headers for line sensor mount
- Ribbon Cable, 11-way or wider, cut with a sharp knife as follows:
  - One 10 inch length of 5 conductor ribbon cable for line sensors

— Two 10 inch lengths of 3 conductor ribbon cable for edge sensors

- Optional: 3 way 0.1" female header for optional charging circuit
- Optional: 3 way 0.1" female header for optional wireless connection

Construction Techniques

This section provides an overview of the motor controller shield construction.

Soldering

Soldering is easy to do if you understand the basic principles and have a little practice. The trick for making a good solder joint is to provide the right amount of heat to the parts to be soldered and use the right solder. 22 AWG solder (0.6mm or .025 inch) or thinner is a good choice for soldering printed circuit boards. A 25-watt to 40-watt iron, ideally with temperature control, is best. The components to be joined should be mechanically secure so they don't move while the solder is cooling—wires should be crimped around terminals (see Figure 4-11 and Figure 4-12). To make the joint, the tip of the iron should have good contact with all the components to be soldered. Feed a small amount of solder where the iron is touching the parts to be joined. When the solder flows around the joint, remove the solder first and then the iron. The connection should be mechanically secure and the joint shiny.

Building the Motor Controller

The motor controller shield is the heart of this robot. As well as controlling the motors, all the sensors are connected to Arduino through this board. The shield is provided as a kit and is the same for use with either the 2WD and 4WD robots, differing only in the method of connecting the motors and mounting to the chassis (both are detailed in later chapters).

The following is an overview of the construction with some tips that you should read through before starting to build the circuit board. You can find step by step construction details for the shield at this site: *http://ladyada.net/make/mshield/solder.html*

Figure 2-1 shows the components for the shield.

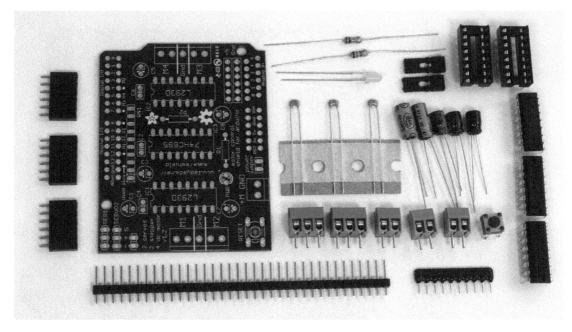

Figure 2-1. *Parts required to build the Motor Shield*

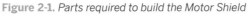

*The parts to the right of (as well as below) the board are packed with the shield, but the three 6-pin headers on the left are not supplied with the standard shield. These headers are used to connect the sensors. These headers **are** included with the Maker Shed companion kits that go along with this book. You can also purchase female headers from Adafruit and other suppliers.*

*The two Maker Shed kits can be found at **http://www.makershed.com/ Bots_and_Bits_for_Bots_s/46.htm**. Look for either the Rovera 2W (Arduino-Controlled 2 Wheel Robotics Platform) or Rovera 4W (Arduino Controlled 4 Wheel Robotics Platform).*

Solder the smallest components first (Figure 2-2). The three small capacitors and two resistors are not polarized so you can insert them either way around.

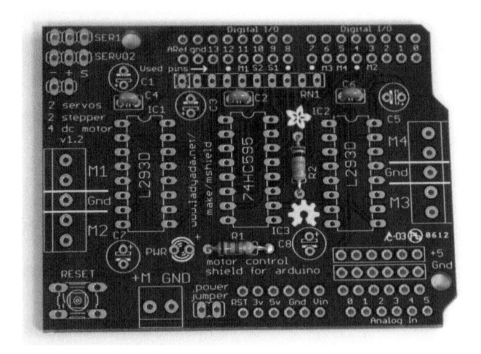

Figure 2-2. *Solder the Small Components*

The resistor network (the long thin component with ten pins) *is* polarized—
the end with the white dot goes to the left of the PCB (nearest to C1) as shown
in Figure 2-3.

Figure 2-3. *Solder the resistor network - the marker (circled) indicates correct orientation*

The large capacitors, ICs, and LED are all polarized. The color of the components
shown in the step-by-step assembly pictures on the Adafruit site (you can find
the link at the beginning of these build notes) may not match the components
or layout for the parts you received (particularly the capacitors) so carefully

check that you have placed the correct value component in the correct orientation. Figure 2-4 shows the layout for version 1.2 of the shield PCB. The kit includes two IC sockets for the L293D chips. As mentioned in the assembly instructions on the Adafruit site, these are optional but if you like to play safe and want to use the sockets, solder them so the indent indicating pin 1 matches the outline printed on the PCB.

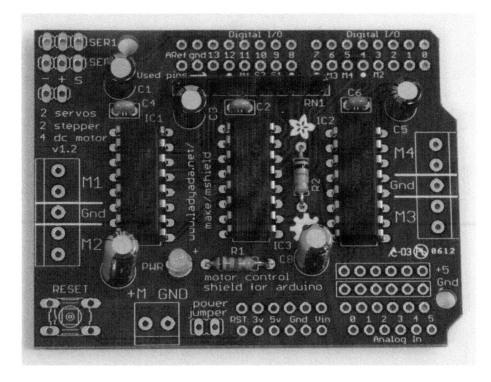

Figure 2-4. *Solder the rest of the polarized components*

Figure 2-5 shows the board with all of the standard shield components (pushbutton, headers, screw terminals) soldered. The final assembly step is to solder the three 6-pin female headers near the analog input pins. These headers are not included in the shield package or mentioned in Adafruit's step-by-step build instructions, but are included with the Maker Shed kits.

Figure 2-5. *Everything soldered except the sensor headers*

Figure 2-6 shows all components including the sensor headers soldered. Trim the component pins (except the header pins that connect the shield to the Arduino) on the underside of the board so they are clear of the Arduino when the shield is plugged onto the board. Locate one of the jumpers supplied with the shield and plug this onto the pins marked *power jumper*—this connects the motor power input and the Arduino VIN (power input) together so both are fed from the batteries that you will be wiring after you have built the robot chassis.

Figure 2-6. *Shield with sensor headers*

Figure 2-7 shows where all of the sensors and other external devices will be connected. The three pin female headers are not needed for some of the projects but you will find it convenient to solder these to the shield at this time.

Figure 2-8 shows two styles of connections. On the left, you'll find the stripboard-based wiring scheme as described in "Making a Line Sensor Mount" (page 17). As you'll see in later chapters, you can experiment with a variety of mounting methods, including the stripboard-based one. The right side of Figure 2-8 shows the wiring for separately connected sensors. As you read through the later chapters and experiment with various mounting techniques, you'll use one or the other wiring schemes. Because you'll be using sockets and ribbon connectors to hook up the sensors, you won't be locked into any particular connection scheme; you can mix and match.

The left and right designation in the diagram refers to left and right from the robot's perspective, and the later chapters will explain where to connect these.

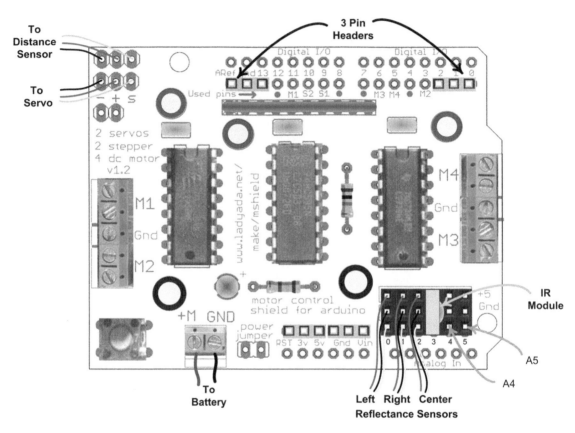

Figure 2-7. *Connections for devices covered in the chapters to come*

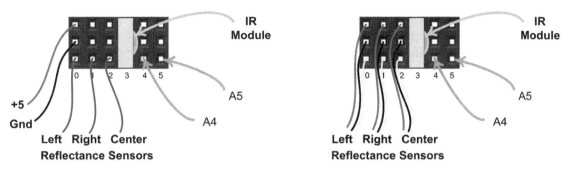

Figure 2-8. *Connection detail - stripboard wiring is shown on the left, individual jumpers shown on the right*

Soldering the Reflectance Sensors

Each sensor package contains a small PCB and a 3-pin header. Insert the header so the shorter length pins emerge on the side of the board with components already soldered, see Figure 2-9. After ensuring you have the header the right way around, solder the three pins.

Figure 2-9. *QTR-1A Reflectance Sensors*

Making a Line Sensor Mount

The line sensing project in this book uses three reflectance sensors wired to analog inputs. Although it is possible to wire the three connections (+5V, Gnd, and Signal) using 9 jumpers, it is more convenient to use a small piece of stripboard to connect the power lines together. Header sockets soldered to the stripboard enable the sensor to be easily unplugged so you can change configuration if you want to swap back and forth between line and edge detection. Figure 2-10 shows the layout of the stripboard (note the five holes you'll need to drill out with a hand drill). Figure 2-12 shows the wires soldered directly to the stripboard pads. If you'd like to add some strain relief, you can drill out a few extra holes in an unused area of the stripboard. Next, divide the wire into two groups (one for positive and negative, and three for the analog pins), and feed the wires through large holes in the board before you solder them. That way, if you tug on the wires, they'll pull against the holes before they pull against your solder joints.

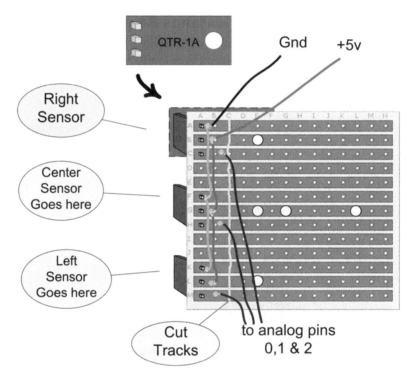

Figure 2-10. *Stripboard layout for mounting QTR-1A reflectance sensors for line following*

To ensure that the mounting bolts don't short the tracks, you can either cut the tracks as shown in Figure 2-10 (you will be cutting along the third column from the left, or the "C" column) or use insulated washers between the bolts heads and the tracks. Figure 2-11 shows how the header sockets are connected, and Figure 2-12 shows the completed stripboard, with the ribbon cable connected. A ten inch length of cable is more than ample. Figure 2-13 shows the other end of the ribbon connected to shield pins.

Figure 2-11. *Stripboard with three 3 pin header sockets*

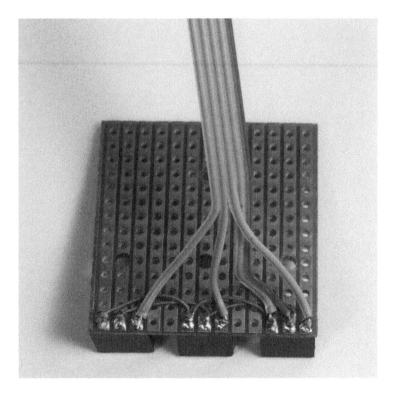

Figure 2-12. *Stripboard with all wires soldered*

Figure 2-13. *Ribbon cable connections to shield pins*

The method of mounting the stripboard depends on the robot chassis; see Chapter 3, *Building the Two-Wheeled Mobile Platform* or Chapter 4, *Building the Four-Wheeled Mobile Platform*. The three holes shown will suit either chassis but you may prefer to wait until you have built the chassis and only drill the holes you need.

Next Steps

The next stage in building the robot is to assemble the chassis. Chapter 3 covers the two-wheeled robot and Chapter 4 is for the four-wheeled version.

Building the Two-Wheeled Mobile Platform

3

This chapter provides advice on the construction of a Two Wheel Drive (2WD) chassis with front caster, as shown in Figure 3-1. Construction is straightforward; you can follow the detailed steps or improvise if you want to customize your robot. The chapter also shows how you attach and connect sensors used in the projects covered in later chapters.

If you prefer to build a two wheeled robot of your own design, you should read the sections on attaching control electronics and sensors; this will prepare you to use the code for the projects in the chapters to come. Information in this chapter my also provide some ideas to help with the design of your own robot.

Figure 3-1. *2WD Robot Chassis*

Hardware Required

See *http://shop.oreilly.com/product/0636920028024.do* for a detailed parts list.

- Tools listed in "Tools" (page 6)
- The assembled electronics (see Chapter 2, *Building the Electronics*
- 2WD Mobile Platform (two wheeled robot kit made by DFRobot)
- Two 0.1uF ceramic capacitors
- Two lengths of 3 conductor ribbon cable, two 3 way 0.1" headers for edge sensors
- Optional: charging circuit resistors and diode, see detailed parts list

Mechanical Assembly

Lay Out the Chassis Parts

Figure 3-2 shows all of the parts contained in the 2WD chassis package. The three black brackets to the left of the figure are not needed for any of the projects in this book.

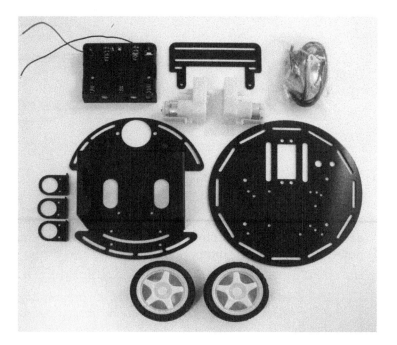

Figure 3-2. *2WD Chassis Parts*

Figure 3-3 shows the contents of the bag containing the mounting hardware. Locate the two bolts with the flat heads and put them aside for mounting the battery case. Also identify the two thicker (M4) bolts that will be used to attach the caster. The remaining short bolts in this pack are identical.

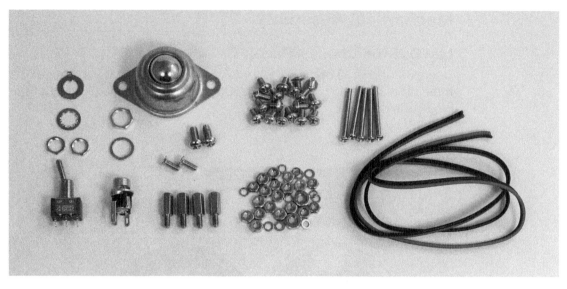

Figure 3-3. *2WD hardware pack contents*

Motor Assembly

Use two long bolts with lock washers and nuts, as shown in Figure 3-5, to attach each motor to the chassis lower plate. Tighten the nuts snugly but take care not to stress the plastic motor housing.

Lock washers are used to prevent a nut from accidentally coming lose due to vibration. This is particularly important for attaching the motor and switch. These washers have a split ring or serrations that apply extra friction when tightened.

If you find that things still come lose, don't overtighten the nuts; an effective solution is retighten the nut and apply a dab of nail polish to the point where the threads emerge from the nut.

Figure 3-4 shows the motors in place with the nut seen on the upper right ready to be tightened.

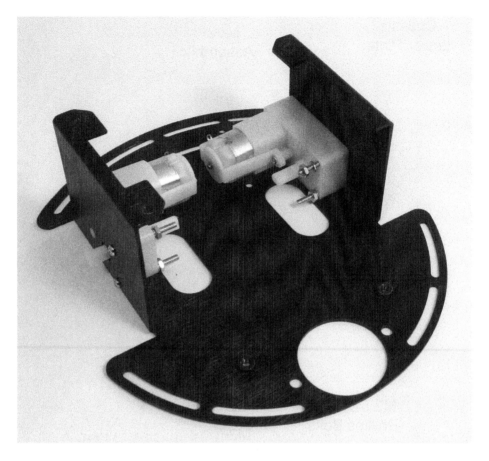

Figure 3-4. *Motors mounted on the chassis lower plate*

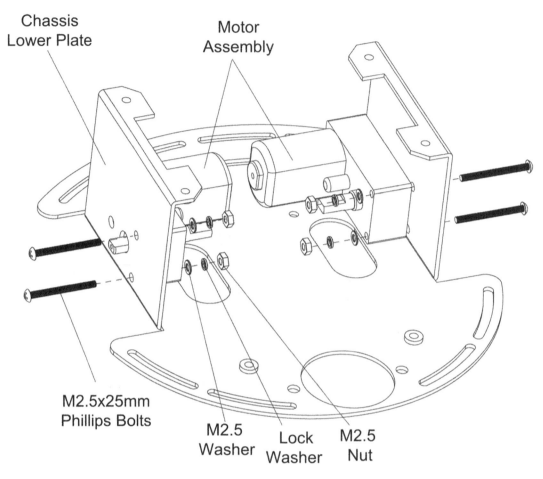

Chassis
Lower Plate

Motor
Assembly

M2.5x25mm
Phillips Bolts

M2.5
Washer

Lock
Washer

M2.5
Nut

Figure 3-5. *Motor Assembly*

Assemble the Chassis Components

Push the wheels onto the motor assembly shafts, aligning the slots in the wheels with the flat section of the motor shaft. Attach the caster with two M4 bolts and nuts. Figure 3-6 and Figure 3-7 show this.

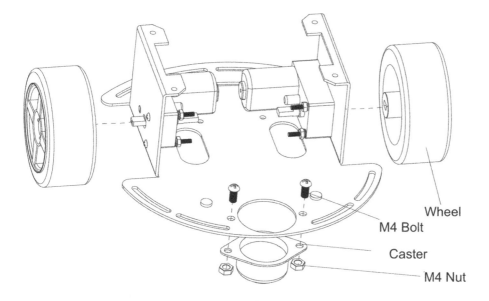

Wheel

M4 Bolt

Caster

M4 Nut

Figure 3-6. *Motor Assembly*

Figure 3-7. *Wheels and caster mounted*

Attach the sensor bracket to the underside of the lower chassis plate, as seen in Figure 3-8 and Figure 3-9.

This robot is sometimes built with the sensor plate mounted at the opposite end of the chassis (furthest from the caster). You can build yours however you like, but the orientation shown here enables the servo mounted distance scanner to be attached in the front of the robot. Also, the sensor bracket in this location maximizes the distance between the wheels and the line sensors and this improves line following sensitivity.

Figure 3-9 shows the underside of the chassis after mounting the sensor bracket. Note that the sensor bracket is attached to the bottom of the chassis plate.

Figure 3-8. *Sensor bracket viewed with the robot right side up*

Figure 3-9. *Sensor bracket viewed with the robot upside down*

The battery pack is bolted to the bottom base plate with two countersunk (flat headed) Phillips bolts as shown in Figure 3-10 and Figure 3-11. You may want to delay this step until after the battery leads have been soldered to make it easier to position all the wires.

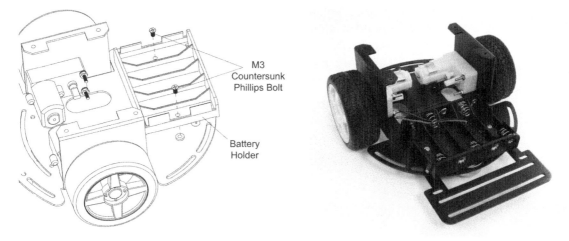

Figure 3-10. *Motor Assembly*

Figure 3-11. *Chassis with Battery Pack Attached*

Cut two pieces of red/black wire, each about 7 1/2 inches long. Strip to expose about 3/16 inch of bare wire at one end of the wires and attach to the motor terminals. Strip 1/4 inch off the other end of the pairs of wires; these will be connected to the motor shield. Connect a 0.1uF capacitor across each of the motor terminals, as shown in Figure 3-12. The capacitors suppress electrical spikes generated by the motor that could interfere with signals on the Arduino board.

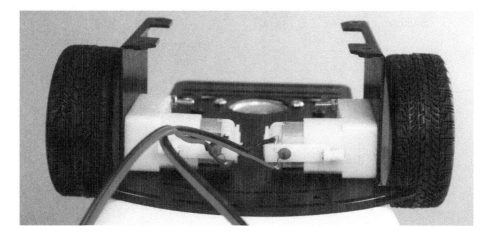

Figure 3-12. *Wires and capacitors soldered to Motors*

The DC power jack is bolted to the top plate using the large (M8) lock washer and nut. The switch is mounted using two (M6) nuts and a lock washer. Put one nut on the switch leaving around 3/16″ of thread above the nut. Then place the lock washer on the thread and push this through the opening in the rear plate and secure with the second M6 nut.

Orient the switch so the toggle moves towards the jack, as shown in Figure 3-13 and Figure 3-14 (Figure 3-15 shows the view from beneath).

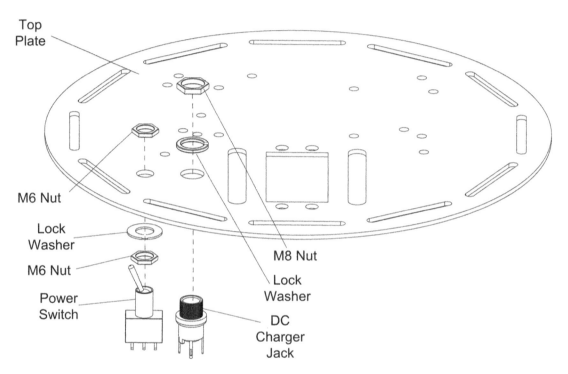

Figure 3-13. *Switch and Jack Assembly*

Figure 3-14. *Top panel showing location of switch and DC jack*

Figure 3-15. *Top panel underside showing orientation of switch and jack*

The battery can be wired as shown in Figure 3-16 and Figure 3-17. The power switch will disconnect the battery when the robot is not in use. The DC jack is not used in this configuration (other than as a junction point for the black ground wires). The switch is off when the toggle is closer to the DC jack as shown (the toggle is a lever; when the exposed end is up as seen in the figure, the contact at the bottom is connected and the contact wired to the shield is open).

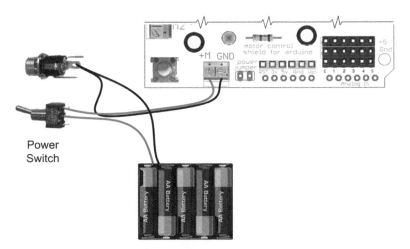

Power Switch

Figure 3-16. *Basic Switch Wiring (no trickle charger)*

Figure 3-17. *Solder the battery wires to the switch*

You can build a simple trickle charger into the robot if you will be using rechargeable NiMH batteries. The charger can be built using the circuit shown in Figure 3-18 and Figure 3-19. See "Trickle Charging" (page 229) for information about using the charger.

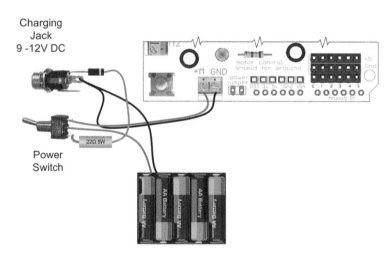

Figure 3-18. *Optional Trickle Charger Wiring*

Figure 3-19. *Wiring of the optional charger jack*

The easiest way to mount the Arduino board is with a strip of Velcro. A 2.5" x 1.5" strip is supplied with the Rovera 2W (Arduino-Controlled 2 Wheel Robotics Platform) kit. To prevent the Arduino pins from accidentally shorting to the chassis, apply insulating tape to the underside of the Arduino board. Gaffer tape works well but you can use (non-conductive) duct tape or heavy duty electrical tape. Attach the 'hairy' side of the Velcro to the taped Arduino board, the hook side is fastened as shown in Figure 3-20.

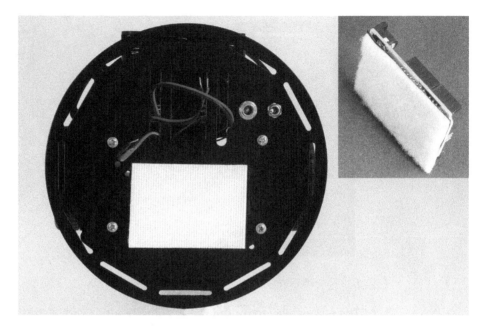

Figure 3-20. *Velcro pad in position on the 2WD chassis. Inset shows Velcro attached to the Arduino board.*

Figure 3-21 shows the mounted boards. The Velcro will hold the boards in position when the robot is moving about, but use one hand to steady the Arduino when you unplug the shield and take care not to use too much downward pressure that could push the Arduino pins through the tape when plugging in the shield.

Figure 3-21. *Arduino board mounted using Velcro*

If you prefer a more rigid mount, you can use two of the 10mm brass standoffs supplied with the chassis and two M3 bolts and nuts (seen on the right side of the board as shown in Figure 3-22). Use a 10mm spacer and M2.5 in the hole near the reset switch. (The hole near the DC jack at the lower left is not used.)

The spacer is required for a Leonardo board because there is insufficient space for an M3 bolt in the munting hole near the switch. The Uno board has more room so you can use a another of the 10mm spacers and M3 hardware for mounting that board.

Figure 3-23 shows the location of the mounting points viewed from the underside of the panel.

Figure 3-22. *Mounting the Arduino board as viewed from the top of the chassis*

Figure 3-23. *Underside showing location of the three Arduino mounting points*

Attach the top plate with four M3 bolts as shown in Figure 3-24.

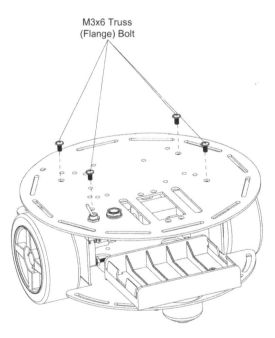

M3x6 Truss
(Flange) Bolt

Figure 3-24. *Top Plate Assembly*

Attaching the Control Electronics

Figure 3-25 shows where the battery and motor wires are connected. Left and right are from the robot's perspective (the right wheel is the one closest to the switch). Figure 3-26 shows the main electronics in place.

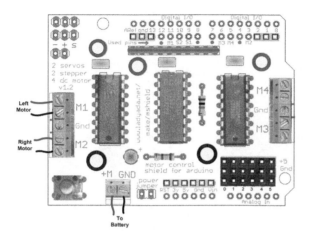

Figure 3-25. *Motor and battery connections*

Figure 3-26. *2WD built and ready to mount sensors*

Mounting the IR sensors

This section covers mounting of the infrared (IR) reflectance sensors for use in edge detecting or line following. "Infrared Reflectance Sensors" (page 134) explains how these sensors work and Chapter 9, *Modifying the Robot to React to Edges and Lines* describes how to use IR sensors. This section explains how to mount these to the 2WD platform and connect them to Arduino. The first projects in this book should have the sensors mounted as shown in the section on edge detection. When you are ready to implement the line following application in Chapter 9, refer back to the section here on positioning the sensors for line following. The stripboard mount described in "Making a Line Sensor Mount" (page 17) simplifies the attachment and wiring of the sensors for line detection and this can also be used for edge detection, but bear in mind that the robot will perform the edge detection task best with the sensors further apart. If the sensors are close together, the robot can have difficulty determining the best angle to turn when an edge is encountered.

Mounting the IR Sensors for Edge Detection

Edge detection requires two QTR-1A sensors mounted on the front of the robot. These should be spaced as widely as possible. The ideal location is with each sensor positioned in front of a wheel so an edge can be detected before a wheel would otherwise fall off a 'cliff'. However, if your priority is simplicity of construction rather than accuracy of edge detection, you can use the same mount described in the next section covering line detection.

The side with the sensor faces the ground and the header pins face upwards. Mount each sensor using a 2-56 bolt and nut (M2 bolts and nuts can also be used) with a 1/2" plastic spacer so the face of the sensor is 3/8" or closer to the ground. Figure 3-27, Figure 3-28, Figure 3-29, and Figure 3-30 show suggested mounting.

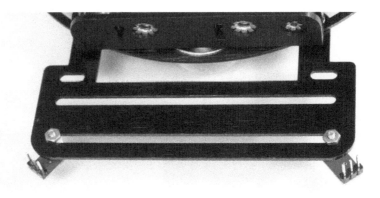

Figure 3-27. Reflectance Sensor location for Edge Detection

Figure 3-28. *Edge Detection Sensor Mounting Detail*

Figure 3-29. *Front view showing location of the Edge Detection Sensors*

Figure 3-30. *Edge sensors wired and ready to run*

Mounting the IR Sensors for Line Following

Three QTR-1a sensors are required for line following. "Making a Line Sensor Mount" (page 17) describes how to build a stripboard mount for line sensing. However, you can also mount and attach each sensors as described in this section if you want to experiment with how varying the spacing of the sensors affects line following.

The sensors can be attached using 2-56 or M2 hardware. The component side faces down and the header pins face upwards. They are mounted in the front, equally spaced with approximately 1/2 inch between the center and the left and right bolts (Figure 3-31).

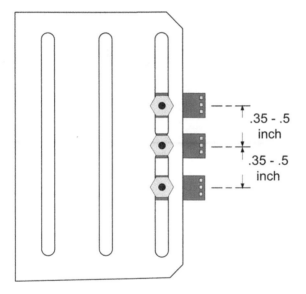

Figure 3-31. *Reflectance Sensor location for Line Following*

If you use the stripboard mount for line sensors covered in Chapter 2, *Building the Electronics*, the stripboard can be mounted above or below the sensor bracket, enabling you to experiment with sensor distance to the ground—but use insulated washers to ensure that the tracks with sensor connections are not shorted to the bracket. Figure 3-32 and Figure 3-33 show how the strip-board can be mounted.

Figure 3-32. *Reflectance Sensor location for line following*

Figure 3-33. *Reflectance Sensor location for line following, alternate view*

See Figure 3-27 for information on connecting the stripboard wires to the motor shield.

Next Steps

Chapter 5, *Tutorial: Getting Started with Arduino* explains how to set up and use the development environment that will be used to upload code to the robot. If you are already an Arduino expert, you can skip to Chapter 6, *Testing the Robot's Basic Functions*, but first, see "Installing Third-Party Libraries" (page 83) for advice on the libraries used with the code for this book.

If you have the libraries installed and want run a simple test to verify that the motors are working correctly, you can run the sketch shown in Example 3-1.

Example 3-1. **Initial motor test for 2WD**

```
/*******************************************
 * MotorTest2wd.ino
 * Initial motor test for 2WD - robot rotates clockwise
 * Left motor driven forward, right backward
 * then counter-clockwise
```

```
 * Michael Margolis 24 July 2012
 ********************************************/
const int LED_PIN = 13;
const int speed = 60; // percent of maximum speed

#include <AFMotor.h>  // adafruit motor shield library (modified my mm)
AF_DCMotor Motor_Left(1, MOTOR12_1KHZ);   // Motor 1
AF_DCMotor Motor_Right(2, MOTOR12_1KHZ);  // Motor 2

int pwm;

void setup()
{
  Serial.begin(9600);
  blinkNumber(8); // open port while flashing. Needed for Leonardo only

  // scale percent into pwm range (0-255)
  pwm= map(speed, 0,100, 0,255);
  Motor_Left.setSpeed(pwm);
  Motor_Right.setSpeed(pwm);
}

// run over and over
void loop()
{
  Serial.println("rotate cw");
  Motor_Left.run(FORWARD);
  Motor_Right.run(BACKWARD);
  delay(5000); // run for 5 seconds
  Serial.println("rotate ccw");
  Motor_Left.run(RELEASE);   // stop the motors
  Motor_Right.run(RELEASE);
  delay(5000); // stop for 5 seconds
}

// function to indicate numbers by flashing the built-in LED
void blinkNumber( byte number) {
   pinMode(LED_PIN, OUTPUT); // enable the LED pin for output
   while(number--) {
     digitalWrite(LED_PIN, HIGH); delay(100);
     digitalWrite(LED_PIN, LOW);  delay(400);
   }
}
```

This sketch runs the motors in opposite directions to cause the robot to rotate clockwise for 5 seconds, then reverses direction to rotate counter-clockwise. This will repeat until the power is switched off.

Building the Four-Wheeled Mobile Platform

4.

This chapter provides advice on the construction of the 4WD (4 Wheel Drive) chassis shown in Figure 4-1. Construction is straightforward—you can follow the detailed steps or improvise if you want to customize your robot. The chapter also shows how you attach and connect sensors used in the projects covered in later chapters.

If you prefer to build a four wheeled robot of your own design, you should read the sections on attaching control electronics and sensors if you want to use the code for the projects in the chapters to come. Information in this chapter my also provide some ideas to help with the design of your own robot.

Figure 4-1. *The 4WD robot chassis*

You will need a Phillips screwdriver, long-nose pliers, wire cutters, wire strippers, a soldering iron, and solder. If you don't have these on hand, you can find more information in Chapter 1, *Introduction to Robot Building*.

Hardware Required

See *http://shop.oreilly.com/product/0636920028024.do* for a detailed parts list.

- Tools listed in "Tools" (page 6)
- The assembled electronics (see Chapter 2, *Building the Electronics*
- 4WD Mobile Platform (four wheeled robot kit made by DFRobot)
- Four 0.1uF ceramic capacitors
- Two lengths of 3 conductor ribbon cable, two 3 way 0.1" headers for edge sensors
- Optional: charging circuit resistors and diode, see detailed parts list

Mechanical Assembly

Mechanical assembly of the 4WD chassis is straightforward and the only tools needed are a Phillips screwdriver and pliers. Following the steps in order will ensure that you use the correct hardware in each assembly. You will need a soldering iron, wire cutters, and wire strippers to wire up the motor and power leads.

Lay Out the Chassis Parts

Figure 4-2 shows all of the parts contained in the 4WD chassis package. Figure 4-3 shows the contents of the bag containing the mounting hardware. The three black brackets to the left of the figure are not needed for any of the projects in this book. Locate the two bolts with the flat heads and put them aside for mounting the battery case. The remaining short bolts in this pack are identical.

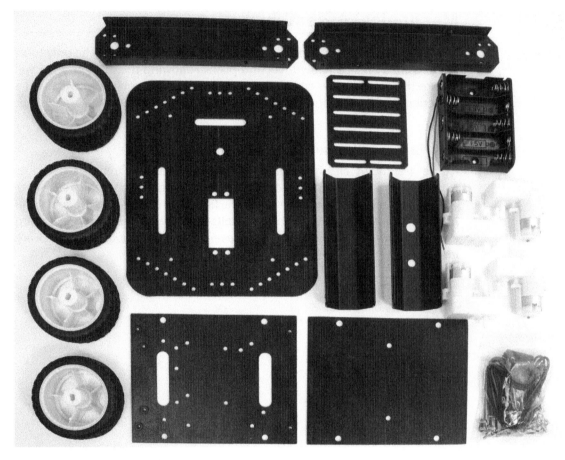

Figure 4-2. *4WD chassis parts*

Figure 4-3. *4WD hardware pack contents*

Motor Assembly

Use four long bolts to attach two motors to each of the side plates. The motor shaft goes through the large hole and there is a small locating stud on the motor that fits into the smaller hole. The lock washer (the one with a raised edge) goes between the nut and flat washer. Ensure the motor is flat against the plate and tighten the nuts firmly but take care not to use too much force or you will stress the plastic motor housing. Figure 4-4 and Figure 4-5 shows the assembly.

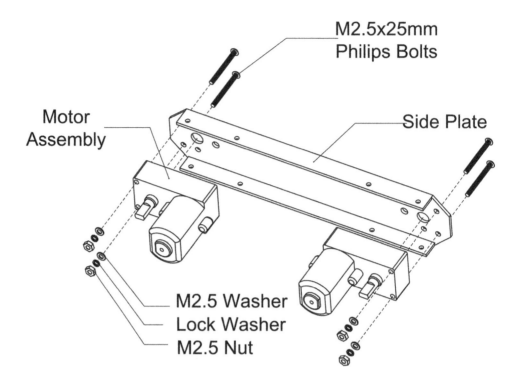

M2.5x25mm
Philips Bolts

Motor
Assembly

Side Plate

M2.5 Washer
Lock Washer
M2.5 Nut

Figure 4-4. *Motor assembly*

Figure 4-5. *Motors mounted onto side plates*

Lock washers are used to prevent a nut from accidentally coming loose due to vibration. This is critical for attaching the motor and switch. These washers have a split ring or serrations that apply extra friction when tightened. If you find that things still come loose, don't overtighten the nuts. Instead, retighten the nut and apply a dab of nail polish to the point where the threads emerge from the nut.

Assemble the Chassis Components

The battery pack is bolted to the bottom base plate with two countersunk (flat headed) Phillips bolts as shown in Figure 4-6 and Figure 4-7.

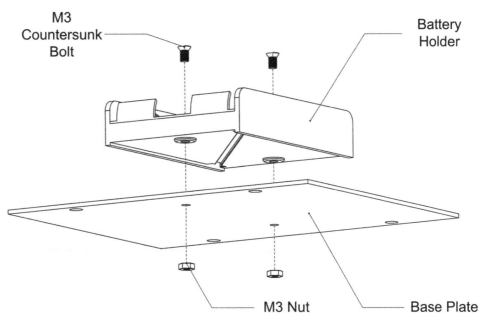

Figure 4-6. *Battery holder assembly*

Figure 4-7. *Battery holder assembly*

The DC power jack is bolted to the rear plate using the large (M8) lock washer and nut as shown in Figure 4-8. The switch is mounted using two nuts and a lock washer (the locating washer is not used). Put one nut on the switch, leaving about enough thread for the nut to be attached to the other side. Place the lock washer on the thread and push this through the opening in the rear plate and secure with the second M6 nut. Orient the switch so the toggle moves from side to side, as shown in the figure. Figure 4-9 and Figure 4-10 show two views of the assembly.

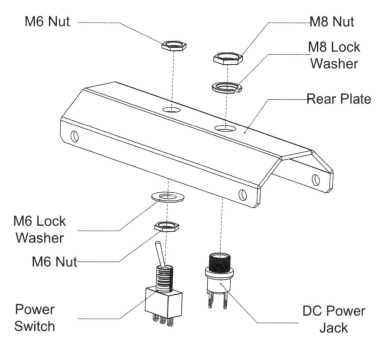

M6 Nut

M8 Nut

M8 Lock Washer

Rear Plate

M6 Lock Washer

M6 Nut

Power Switch

DC Power Jack

Figure 4-8. *Switch and power jack assembly*

Figure 4-9. *Rear panel switch and power jack assembly viewed from the front*

Figure 4-10. *The other side of the panel showing the switch orientation and power jack*

Solder the Power and Motor Connections

It is easier to solder the connections before everything is bolted together. The motor connections use the red and black wire provided in the kit. Cut four pieces, each three inches long. Strip 1/4 inch off the red and black wires on one end; this end connects to the motor shield. The other end is connected to the motor terminals; strip to expose about 3/16 inch of bare wire. Connect a 0.1uF capacitor across each of the motor terminals, as shown in Figure 4-11. The capacitors suppress electrical spikes generated by the motor that could interfere with signals on the Arduino board. Connect and crimp the wires as shown in Figure 4-12, and then solder the wires and capacitors to the motor terminals as shown in Figure 4-13.

Figure 4-11. *Crimp the capacitor leads to the motor terminals*

Figure 4-12. *Crimp the wires*

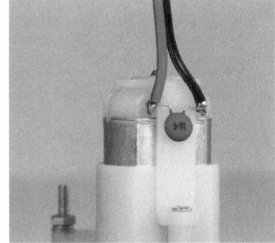

Figure 4-13. *Solder the motor terminals*

Connecting the Battery Pack and Power Switch

The battery can be wired as shown in Figure 4-14, but you cannot charge the battery in this configuration. The power switch will disconnect the battery when the robot is not in use. The DC jack is not used in this configuration (other than as a junction point for the black ground wires). The switch is off when the toggle is closer to the DC jack as shown (the toggle is a lever, when the exposed end is up as seen in the figure, the contact at the bottom is connected and the contact wired to the shield is open). Figure 4-15 shows the completed circuit.

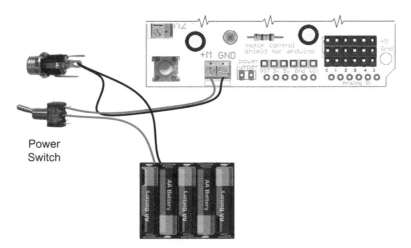

Figure 4-14. Basic switch wiring (no trickle charger)

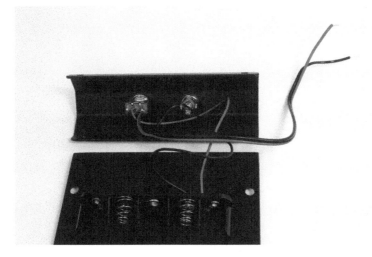

Figure 4-15. Red wires soldered to switch

Building the Optional Trickle Charger

You can build a simple trickle charger into the robot if you will be using rechargeable NiMH batteries. See "Trickle Charging" (page 229) for information about using the charger.

The circuit is wired as shown in Figure 4-16 and Figure 4-17. The battery is connected to both the robot and charger when it is switched on, enabling the Arduino to monitor and display the battery voltage. The connection via the resistor to pin 13 is required to tell the Arduino that a charger is connected so it can monitor the voltage instead of driving the robot.

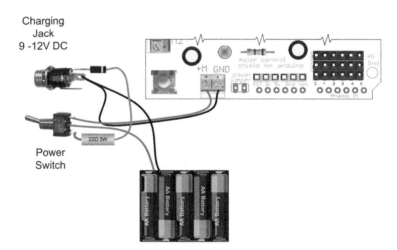

Figure 4-16. *Wiring for trickle charging with Arduino voltage monitoring*

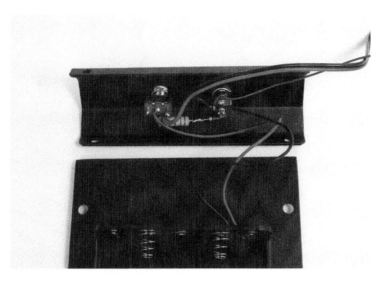

Figure 4-17. *Wiring for trickle charging with Arduino voltage monitoring*

Assemble the Chassis

Attach the front and rear plates to the sides using eight of the M3x6 bolts (Figure 4-18). The sides are symmetrical so it doesn't matter which end goes to the front or back.

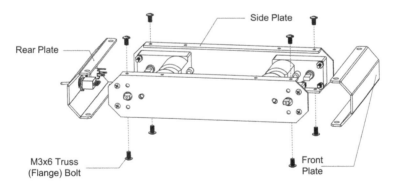

Figure 4-18. *Chassis assembly*

Attach the bottom plate using four M3x6 bolts (Figure 4-19).

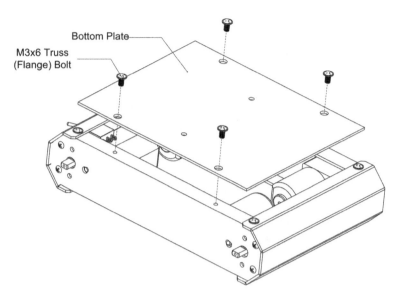

Bottom Plate

M3x6 Truss
(Flange) Bolt

Figure 4-19. *Bottom plate assembly*

Mounting Arduino and Connecting Wires to the Shield

The easiest way to mount the Arduino board is with a strip of Velcro. A 2.5" x 1.5" strip is supplied with the Rovera 4W (Arduino-Controlled 4 Wheel Robotics Platform) kit. To prevent the Arduino pins from accidentally shorting to the chassis, apply insulating tape to the underside of the Arduino board. Gaffer tape works well but you can use (non-conductive) duct tape or heavy duty electrical tape. Attach the 'hairy' side of the Velcro to the taped Arduino board; the hook side is fastened as shown in Figure 4-20. Figure 4-21 shows some other views of this.

Figure 4-20. *Velcro pad in position on the top plate*

The Velcro will hold the boards in position when the robot is moving about, but use one hand to steady the Arduino when you unplug the shield and take care not to use too much downward pressure that could push the Arduino pins through the tape when plugging in the shield.

Figure 4-21. *Inset shows Velcro attached to the Arduino board.*

If you prefer a more rigid mount, you can use three 3/8" or 1/4 inch (5mm) spacers with three 1/2 inch 2-56 bolts and nuts. Figure 4-22 and Figure 4-23 show the location of the mounting hardware.

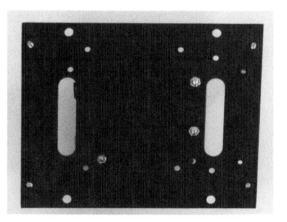

Figure 4-22. *Arduino board mounted using three spacers*

Figure 4-23. *Underside view showing mounting nuts*

Figure 4-24 shows the motor wires and battery wires inserted through the cutouts in the top plate ready for the connections shown in Figure 4-25 .

Figure 4-24. *Wires ready to connect to shield*

Figure 4-25. *Wires connected*

Figure 4-26 shows how the motor and battery wires attach to the connectors on the motor shield.

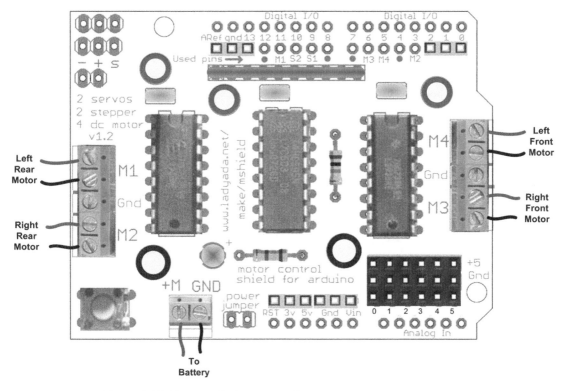

Figure 4-26. *Motor and battery connections*

Attach the sensor plate with two M3 bolts as shown in Figure 4-27; the top plate is attached using four M3 bolts as seen in Figure 4-28.

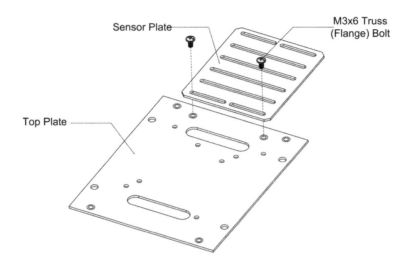

Figure 4-27. *Sensor plate assembly*

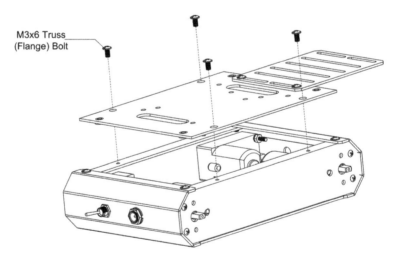

Figure 4-28. *Top plate assembly*

The upper deck is bolted to four 50mm standoffs that are attached as shown in Figure 4-29.

M3x6 Truss (Flange) Bolt

M2x50 Standoff

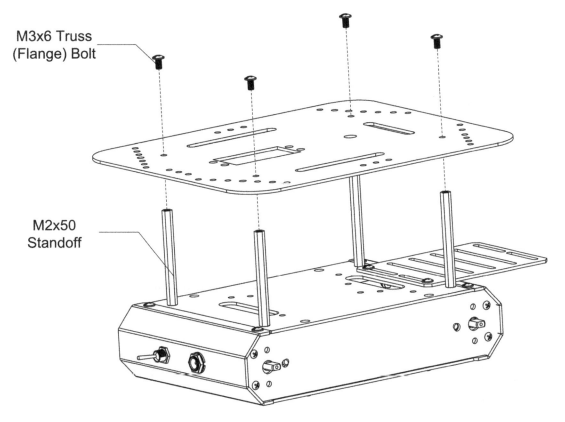

Figure 4-29. *Attach the upper deck*

Figure 4-30 shows the fully-assembled chassis (a side view is visible in Figure 4-31). You can see the front and rear views in Figure 4-32 and Figure 4-33.

Figure 4-30. *The assembled chassis*

Figure 4-31. *Side view*

Figure 4-32. *Front view*

Figure 4-33. *Rear view*

Mounting the IR sensors

This section covers mounting of the infrared (IR) reflectance sensors for use in edge detecting or line following. "Infrared Reflectance Sensors" (page 134) explains how these sensors work and Chapter 9, *Modifying the Robot to React to Edges and Lines* describes how to use IR sensors. This section explains how to mount these sensors onto the 4WD platform and connect them to Arduino. The first projects in this book should have the sensors mounted as shown in the section on edge detection. When you are ready to implement the line following application in Chapter 9, refer back to the section on positioning the sensors for line following. The stripboard mount described in "Making a Line Sensor Mount" (page 17) simplifies the attachment and wiring of the sensors for line detection and this can also be used for edge detection, but bear in mind that the robot will perform the edge detection task best with the sensors further apart. If the sensors are close together, the robot can have difficulty determining the best angle to turn when an edge is encountered.

Mounting the IR Sensors for Edge Detection

Edge detection requires two QTR-1A sensors mounted on the front of the robot. These should be spaced as widely apart as possible. The ideal location is with each sensor positioned in front of a wheel so an edge can be detected before a wheel would otherwise fall off a 'cliff'. However, if your priority is simplicity of construction rather than accuracy of edge detection, you can use the same mount described in the next section covering line detection. But bear in mind that the robot will perform the edge detection task best with the sensors further apart. If the sensors are close together, the robot can have difficulty determining the best angle to turn when an edge is encountered.

Mount each sensor using a 2-56 bolt and nut (M2 bolt and nut can also be used). The component side faces the ground and the header pins face upwards. The sensors can be angled as shown in Figure 4-34 and Figure 4-35.

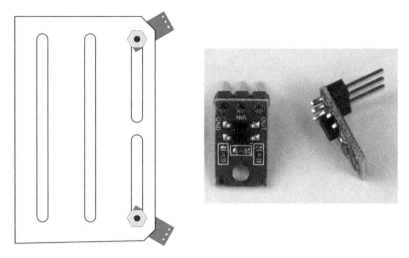

Figure 4-34. *Reflectance sensor location for edge detection*

Figure 4-35. *Reflectance sensor location for edge detection*

Mounting the IR Sensors for Line Following

Three QTR-1a sensors are required for line following. "Making a Line Sensor Mount" (page 17) describes how to build a stripboard mount for line sensing. However, you can also mount and attach each sensors as described in this section if you want to experiment with how varying the spacing of the sensors affects line following. Like the edge sensors, they can be attached using 2-56 or M2 hardware. The component side faces down and the header pins face upwards. They are mounted in the front, equally spaced with approximately 1/2 inch between the center and the left and right bolts (see Figure 4-36). Figure 4-37 and Figure 4-38 show the sensors attached to the chassis.

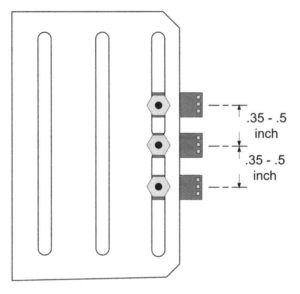

Figure 4-36. *Reflectance sensor location for line following*

Figure 4-37. *Reflectance sensor location for line follow-ing, viewed from front*

Figure **4-38.** *Reflectance sensor location for line fol-lowing, alternate view*

Next Steps

Chapter 5, *Tutorial: Getting Started with Arduino* explains how to set up and use the development environment that will be used to upload code to the robot. If you are already an Arduino expert, you can skip to Chapter 6, *Testing the Robot's Basic Functions*, but first, see "Installing Third-Party Libraries" (page 83) for advice on the libraries used with the code for this book and the steps needed to configure the RobotMotor library for the 4WD robot.

If you have the libraries installed and want run a simple test to verify that the motors are working correctly, you can run the following sketch:

Example 4-1. **Initial motor test for 4WD**

```
/******************************************
 * MotorTest4wd.ino
 * Initial motor test for 4WD
 * robot rotates clockwise
 *  (Left motors driven forward, right backward)

 * Michael Margolis 24 July 2012
 ******************************************/
const int LED_PIN = 13;
const int speed = 60; // percent of maximum speed

#include <AFMotor.h>  // adafruit motor shield library (modified my mm)
AF_DCMotor Motor_Left_Front(4, MOTOR34_1KHZ);    // Motor 4
AF_DCMotor Motor_Right_Front(3, MOTOR34_1KHZ);   // Motor 3
AF_DCMotor Motor_Left_Rear(1, MOTOR12_1KHZ);     // Motor 1
```

```
AF_DCMotor Motor_Right_Rear(2, MOTOR12_1KHZ);    // Motor 2

int pwm;

void setup()
{
  Serial.begin(9600);
  blinkNumber(8); // open port while flashing. Needed for Leonardo only

  // scale percent into pwm range (0-255)
  pwm= map(speed, 0,100, 0,255);
  Motor_Left_Front.setSpeed(pwm);
  Motor_Right_Front.setSpeed(pwm);
  Motor_Left_Rear.setSpeed(pwm);
  Motor_Right_Rear.setSpeed(pwm);
}

// run over and over
void loop()
{
  Serial.println("rotate cw");
  Motor_Left_Front.run(FORWARD);
  Motor_Left_Rear.run(FORWARD);

  Motor_Right_Front.run(BACKWARD);
  Motor_Right_Rear.run(BACKWARD);

  delay(5000); // run for 5 seconds
  Serial.println("stopped");
  Motor_Left_Front.run(RELEASE);   // stop the motors
  Motor_Right_Front.run(RELEASE);
  Motor_Left_Rear.run(RELEASE);   // stop the motors
  Motor_Right_Rear.run(RELEASE);

  delay(5000); // stop for 5 seconds
}

// function to indicate numbers by flashing the built-in LED
void blinkNumber( byte number) {
    pinMode(LED_PIN, OUTPUT); // enable the LED pin for output
    while(number--) {
      digitalWrite(LED_PIN, HIGH); delay(100);
      digitalWrite(LED_PIN, LOW);  delay(400);
    }
}
```

This sketch runs the motors in opposite directions to cause the robot to rotate clockwise for 5 seconds, then stops for 5 seconds. This will repeat until the power is switched off.

*This test sketch does not use the RobotMotor library—if this test functions correctly but the test in **Chapter 6, Testing the Robot's Basic Functions** does not work, the most likely cause is the configuration of the motor library—make sure you copy the 4wd version of the library code as described in **"Installing Third-Party Libraries" (page 83)**.*

Tutorial: Getting Started with Arduino

5

The Arduino environment has been designed to be easy to use for beginners who have no software or electronics experience. If you are new to Arduino, this chapter will help you get started but you will need to consult the Arduino online help and a good book on Arduino will be a big help (the author's "Arduino Cookbook" is highly recommended as reference.)

> *If you're already familiar with Arduino, please feel free to skip the introductory material in this chapter. However, you will need to install the libraries that are included in the download the code available from: http://shop.oreilly.com/product/0636920028024.do. The section "Installing Third-Party Libraries" (page 83) has details on installing the required libraries.*

Arduino is best known for its hardware, but you also need software to program that hardware. Both the hardware and the software are called "Arduino." The combination enables you to create projects that sense and control the physical world. The software is free, open source, and cross-platform. The boards are inexpensive to buy or you can build your own (the hardware designs are also open source). In addition, there is an active and supportive Arduino community that is accessible worldwide through the Arduino forums and the wiki (known as the Arduino Playground). The forums and the wiki offer project development examples and solutions to problems that can provide inspiration and assistance as you pursue your own projects.

The information in this chapter will get you started by explaining how to set up the development environment and how to compile and run an example *sketch*.

> *Source code containing computer instructions for controlling Arduino functionality is usually referred to as a **sketch** in the Arduino community. The word **sketch** will be used throughout this book to refer to Arduino program code.*

The Blink sketch, which comes with Arduino, is used as an example sketch in this chapter. If you have already assembled the robot and downloaded the source code for this book, feel free to use the HelloRobot sketch described in Chapter 6, *Testing the Robot's Basic Functions*.

> *If you don't have the Arduino software and driver installed on your machine, wait until **"Connecting the Arduino Board" (page 78)** to plug the Arduino into your computer.*

Hardware Required

- Computer with Arduino 1.0.1 or later installed
- Leonardo (or Uno) Arduino board
- Motor Shield (see Chapter 2, *Building the Electronics*)
- USB cable

Arduino Software

Software programs, called sketches, are created on a computer using the Arduino *integrated development environment* (IDE). The IDE enables you to write and edit code and convert this code into instructions that Arduino hardware understands. The IDE also transfers those instructions to the Arduino board (a process called *uploading*).

Arduino Hardware

The Arduino board is where the code you write is executed. The board can only control and respond to electricity, so specific components are attached to it to enable it to interact with the real world. These components can be *sensors*, which convert some aspect of the physical world to electricity so that the board can sense it, or *actuators*, which get electricity from the board and convert it into something that changes the world. Examples of sensors include switches, accelerometers, and ultrasound distance sensors. Actuators are things like lights and LEDs, speakers, motors, and displays.

There are a variety of official boards that you can run your Arduino sketches on and a wide range of Arduino-compatible boards produced by members of the community.

The most popular boards contain a USB connector that is used to provide power and connectivity for uploading your software onto the board. Figure 5-1 shows the board used for the robots in this book, the Arduino Leonardo.

Figure 5-1. *Basic board: the Arduino Leonardo*

You can get boards that are smaller and boards with more connections. The Leonardo is used with these robotics projects because it is inexpensive but you can use other boards such as the Uno if you prefer.

*If you want to use an Uno board (or earlier Arduino boards), you may need to use a s slightly higher voltage (an additional battery) to power the robot, see **Appendix D**).*

Add-on boards that plug into Arduino to extend hardware resources are called *shields*. The robots covered in this book use a shield that controls the direction and speed of the motors, see Figure 5-2:

Figure 5-2. *Motor Shield plugged into the Arduino Leonardo*

Online guides for getting started with Arduino are available at *http://ardui no.cc/en/Guide/Windows* for Windows, *http://arduino.cc/en/Guide/MacOSX* for Mac OS X, and *http://www.arduino.cc/playground/Learning/Linux* for Linux.

Installing the Integrated Development Environment (IDE)

The Arduino software for Windows, Mac, and Linux can be downloaded from *http://arduino.cc/en/Main/Software*.

Installing Arduino on Windows

The Windows download is a ZIP file. Unzip the file to any convenient directory —*Program Files/Arduino* is a sensible place.

Unzipping the file will create a folder named *Arduino-1.0.<nn>* (where *<nn>* is the version number of the Arduino release you downloaded). The directory contains the executable file (named *Arduino.exe*), along with various other files and folders. Double-click the *Arduino.exe* file and the splash screen should appear (see Figure 5-3), followed by the main program window (see Figure 5-4). Be patient, as it can take some time for the software to load.

Figure 5-3. *Arduino splash screen (Version 1.0 in Windows 7)*

Installing Arduino on OS X

The Arduino download for the Mac is a disk image (*.dmg*); double-click the file when the download is complete. The image will mount (it will appear like a memory stick on the desktop). Inside the disk image is the Arduino application. Copy this to somewhere convenient—the *Applications* folder is a sensible place. Double-click the application once you have copied it over (it is not a good idea to run it from the disk image). The splash screen will appear, followed by the main program window (Figure 5-4).

Figure 5-4. *IDE main window (Arduino 1.0 on a Mac)*

Installing Arduino on Linux

Linux installation varies depending on the Linux distribution you are using. See the Arduino wiki for information (*http://www.arduino.cc/playground/Learn ing/Linux*).

Driver Installation

To enable the Arduino development environment to communicate with the board, the operating system needs to use the appropriate drivers for your board.

On Windows, use the USB cable to connect your PC and the Arduino board and wait for the Found New Hardware Wizard to appear. If you are using a Leonardo or Uno board let the wizard attempt to find and install drivers.

Troubleshooting the Found New Hardware Wizard

If the Found New Hardware Wizard does not appear when you first connect a Leonardo board, open Device Manager as described in the next paragraph and if you see Other device> Arduino Leonardo with an exclamation point, right click on the entry and select Update Driver Software. Choose the Browse my computer for Driver Software option, and navigate to the Drivers folder inside the Arduino folder you just unzipped. Select the drivers folder and windows should then proceed with the installation process. If the Windows can't verify the publisher of the driver software dialog pops up, select Install this software anyway.

If the Wizard starts but fails to find drivers (don't worry, this is the expected behavior with an Uno board). To fix it you now need to go to Start Menu>Control Panel>System and Security. Click on System, and then open Device Manager. In the listing that is displayed find the entry in COM and LPT named Arduino UNO (COM nn). nn will be the number Windows has assigned to the port created for the board. You will see a warning logo next to this because the appropriate drivers have not yet been assigned. Right click on the entry and select Update Driver Software. Choose the Browse my computer for Driver Software option, and navigate to the Drivers folder inside the Arduino folder you just unzipped. Select the ArduinoUNO.inf file and windows should then proceed with the installation process. If the Windows can't verify the publisher of the driver software dialog pops up, select Install this software anyway.

If you are using an earlier board (any board that uses FTDI drivers) with Windows Vista or Windows 7 and are online, you can let the wizard search for drivers and they will install automatically. On Windows XP (or if you don't have internet access), you should specify the location of the drivers. Use the file selector to navigate to the *FTDI USB Drivers* directory, located in the directory where you unzipped the Arduino files. When this driver has installed, the Found New Hardware Wizard will appear again, saying a new serial port has been found. Follow the same process as before.

On the Mac, the latest Arduino boards can be used without additional drivers. When you first plug the board in a notification will pop up saying a new network port has been found; you can dismiss this. If you are using earlier boards (boards that need FTDI drivers), you will need to install driver software. There is a package named *FTDIUSBSerialDriver*, with a range of numbers after it, inside the Arduino installation disk image. Double-click this and the installer will take you through the process. You will need to know an administrator password to complete the process.

On Linux, most distributions have the driver already installed, but follow the Linux link given in "Arduino Hardware" (page 72) for specific information for your distribution.

*If the software fails to start, check the troubleshooting section of the Arduino website, **http://arduino.cc/en/Guide/Troubleshooting**, for help solving installation problems.*

Connecting the Arduino Board

Plug the board into a USB port on your computer and check that the green LED power indicator on the board illuminates. The location of the LED is indicated in Figure 5-5.

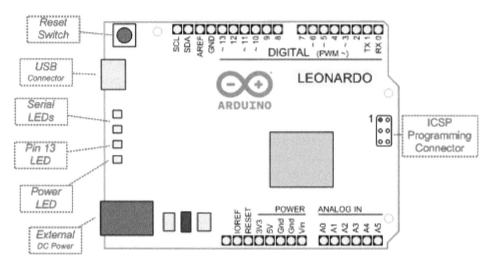

Figure 5-5. *The Leonardo*

If you have a factory fresh board, an orange LED (labeled "Pin 13 LED" in Figure 5-5) should flash on and off when the board is powered up (boards come from the factory preloaded with software to flash the LED as a simple check that the board is working).

If the power LED does not illuminate when the board is connected to your computer, the board is probably not receiving power.

Using the IDE

Use the Arduino IDE to create, open, and modify sketches that define what the board will do. You can use buttons along the top of the IDE to perform these actions (shown in Figure 5-6), or you can use the menus or keyboard shortcuts (some are shown in Figure 5-7).

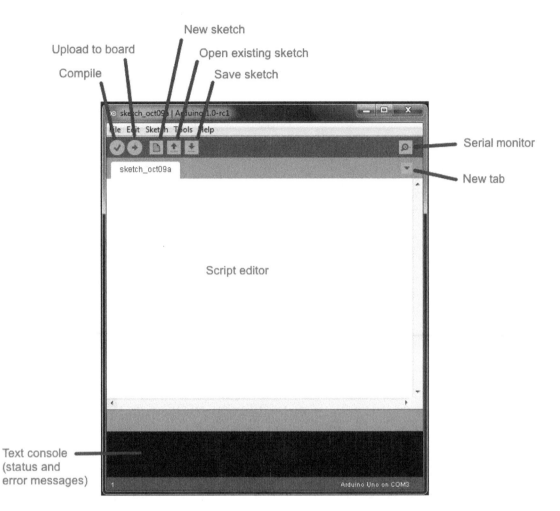

Figure 5-6. *Arduino IDE*

The Sketch Editor area is where you view and edit code for a sketch. It supports common text editing keys such as Ctrl-F (⌘-F on a Mac) for find, Ctrl-Z (⌘-Z on a Mac) for undo, Ctrl-C (⌘-C on a Mac) to copy highlighted text, and Ctrl-V (⌘-V on a Mac) to paste highlighted text.

Figure 5-7 shows how to load the Blink sketch (the sketch that comes preloaded on a new Arduino board).

After you've started the IDE, go to the File→Examples menu and select 1.Basics→Blink, as shown in Figure 5-7. The code for blinking the built-in LED will be displayed in the Sketch Editor window.

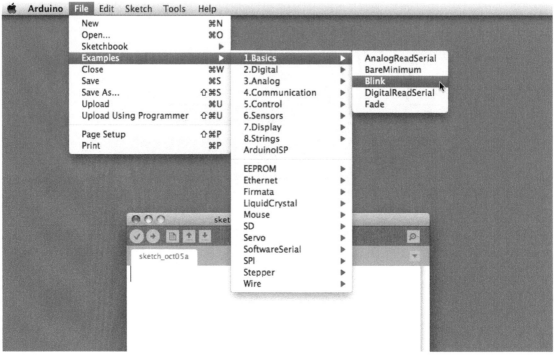

Figure 5-7. *IDE menu (selecting the Blink example sketch)*

Before the code can be sent to the board, it needs to be converted into instructions that can be read and executed by the Arduino controller chip; this is called *compiling*. To do this, click the compile button (the top-left button with a tick inside), or select Sketch→Verify/Compile (Ctrl-R, ⌘-R on a Mac).

You should see a message that reads "Compiling sketch..." and a progress bar in the message area below the text editing window. After a second or two, a message that reads "Done Compiling" will appear. The black console area will contain the following additional message:

```
Binary sketch size: 1026 bytes (of a 32256
byte maximum)
```

The exact message may differ depending on your board and Arduino version; it is telling you the size of the sketch and the maximum size that your board can accept.

The final message telling you the size of the sketch indicates how much program space is needed to store the controller instructions on the board. If the size of the compiled sketch is greater than the available memory on the board, the following error message is displayed:

```
Sketch too big;
see http://www.arduino.cc/en/Guide/Troubleshooting#size for tips
on reducing it.
```

If this happens, you need to make your sketch smaller to be able to put it on the board, or get a board with higher capacity. You will not have this problem with the Blink example sketch.

If there are errors in the code, the compiler will print one or more error messages in the console window. These messages can help identify the error.

As you develop and modify a sketch, you should also consider using the File→Save As menu option and using a different name or version number regularly so that as you implement each bit, you can go back to an older version if you need to.

Uploading and Running the Blink Sketch

To transfer your compiled sketch to the Arduino board, connect your board to your computer using the USB cable. Load the sketch into the IDE as described in "Using the IDE" (page 78).

Next, select Tools→Board from the drop-down menu and select the name of the board you have connected.

Now select Tools→Serial Port. You will get a drop-down list of available serial ports on your computer. Each machine will have a different combination of serial ports, depending on what other devices you have used with your computer.

On Windows, they will be listed as numbered COM entries. If there is only one entry, select it. If there are multiple entries, your board will probably be the last entry.

On the Mac, your board will be listed twice (you can use either one):

```
/dev/tty.usbmodemXXXXXXX
/dev/cu.usbmodemXXXXXXX
```

If you have an older board, it will be listed as follows:

```
/dev/tty.usbserial-XXXXXXX
/dev/cu.usbserial-XXXXXXX
```

Each board will have different values for *XXXXXXX*. Select either entry.

Click on the upload button (in Figure 5-6, it's the second button from the left), or choose File→Upload to I/O board (Ctrl-U, ⌘-U on a Mac).

The software will compile the code, as in "Using the IDE" (page 78). After the software is compiled, it is uploaded to the board. If you look at your board, you will see the LED stop flashing, and two lights (labeled as Serial LEDs in Figure 5-5) just below the previously flashing LED should flicker for a couple of seconds as the code uploads. The original light should then start flashing again as the code runs.

The IDE will display an error message if the upload is not successful. Problems are usually due to the wrong board or serial port being selected or the board not being plugged in. The currently selected board and serial port are displayed in the status bar at the bottom of the Arduino window

If you have trouble identifying the correct port on Windows, try unplugging the board and then selecting Tools→Serial Port to see which COM port is no longer on the display list. Another approach is to select the ports, one by one, until you see the lights on the board flicker to indicate that the code is uploading.

Using Tabs

Tabs provide a convenient way to organize code when your sketch starts to grow. It enables you to keep functionally related code together and simplifies sharing this code across more than one sketch.

The arrow in the upper right of Figure 5-8 points to the button which invokes a drop-down window of tab related functions. This window displays the names of the tabs and offers a list of commands:

- New Tab creates a new tab that (you will be prompted to name the tab)
- Rename enables you to change the name of the currently selected tab
- Delete deletes the current tab (you are asked if you are sure you want to do that)

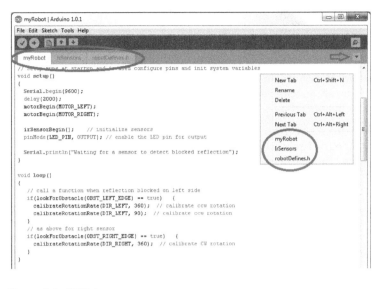

Figure 5-8. *IDE tabs*

Each tab is a separate file and when you copy these files to other sketches you add the tab to that sketch.

Because there are many functional modules used in this book and these are shared across most of the sketches, tabs are used extensively. Figure 5-8 shows the myRobot sketch, discussed in the next chapter, that uses tabs for infrared sensor code (irSensors) and for program constants and definitions (*robotDe fines.h*).

Installing Third-Party Libraries

The download code for this book (see "How to Contact Us" (page xv)) contains three libraries that are required to run all the sketches described in the book. These libraries are in a folder called *libraries* in the zip. You need to copy these so they are in a folder called *libraries* inside your Arduino document folder. To find the Arduino document folder, open Preferences (Arduino→Preferences on Mac; File→Preferences on Windows) and note the sketchbook location. Navigate to that directory in a file system browser (such as Windows Explorer or the OS X Finder) or at the terminal. If no *libraries* folder exists, create one and put the folder you unzipped inside it.

If the Arduino IDE is still running, quit and restart it. The IDE scans this folder to find libraries only when it is launched. If you now go to the menu Sketch→Import Library, at the bottom, below the gray line and the word *Con tributed*, you should see the library you have added.

Configuring the Library for Four Wheels

If your robot uses four wheel drive, you must configure the *RobotMotor* library code by modifying the *RobotMotor.h* file to tell the compiler that the library should be built for the 4WD chassis.

To modify the *RobotMotor.h* file to use the 4WD chassis, first go to Arduino's preferences (File→Preferences on Windows or Linux, Arduino→Preferences on Mac). Under Sketchbook Location, you'll find the name of the directory that contains your sketches and libraries. Next:

1. Open the sketchbook folder in the Finder (Mac) or Explorer (Windows).

2. Locate the libraries directory inside, and then open the directory named *RobotMotor* .

3. Right-click the *RobotMotor.h* file, and open it with a plain text editor. On Windows, you can use Notepad. On the Mac, you can use TextEdit. On Linux, use your favorite plain text editor.

4. Change #define CHASSIS_2WD to #define CHASSIS_4WD and save the file

A version of *RobotMotor.h* with the modification for 4wd preconfigured is in the *libraries/RobotMotor/RobotMotor4wd* folder of the examples zip file. Copying this file to the *RobotMotor* folder will replacing the 2wd with the 4wd version of the file.

If the libraries provide example sketches, you can view these from the IDE menu; click File→Examples, and the library's examples will be under the library's name in a section between the general examples and the Arduino distributed libraries example listing.

If the library examples do not appear in the Examples menu or you get a message saying "Library not found" when you try to use the library, check that the *libraries* folder is in the correct place with the name spelled correctly. A library folder named *<LibraryName>* (where *<LibraryName>* is the name for the library) must contain a file named *<LibraryName>.h* with the same spelling and capitalization. Check that additional files needed by the library are in the folder.

Testing the Robot's Basic Functions

§

In this chapter, you will upload a test sketch to the robot that will verify that your robot is working correctly.

Hardware Required

- The assembled robot chassis.
- Motors connected to shield (see Figure 3-25 for 2WD or Figure 4-26 for 4WD).
- Example code and libraries installed, see "Installing Third-Party Libraries" (page 83).
- 5 AA cells inserted into the battery holder (USB does not provide sufficient power to drive the motors).
- Reflectance sensors mounted and connected (left sensor to analog input 0, right to analog 1). You can use the stripboard wiring described in "Making a Line Sensor Mount" (page 17). But to run the edge detecting project described in Chapter 9, you need more space between the sensors.

Figure 6-1 shows the assembled two wheel robot; Figure 6-2 shows the assembled four wheel robot. Figure 6-3 shows the sensor and motor connections.

Figure 6-1. *Two wheeled robot with reflectance sensors*

Figure 6-2. *Four wheeled robot with reflectance sensors*

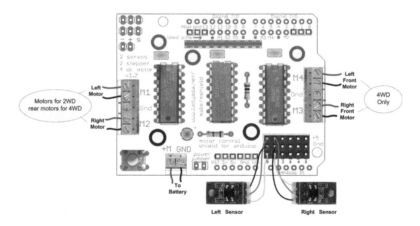

Figure 6-3. *Reflectance sensor connections*

Software Prerequisites

Although the sketch code used in this chapter is printed in the pages that follow, you will need some libraries that are included in example code (see "How to Contact Us" (page xv) for the URL). The sketch folders can be copied to your Arduino sketchbook folder (if you are not familiar with the Arduino environment, read through Chapter 5). The download files in the library folder must be copied to your Arduino libraries folder (see "Installing Third-Party Libraries" (page 83)).

Install the AFMotor library contained in the download zip file. This library is modified from the one on the Adafruit site to work with the Leonardo board; the standard Adafruit library can be used with the Uno board.

Install the RobotMotor library contained in the example code download. This library comes configured for the two wheeled robot; if you have the four wheeled robot will need to update the library for this robot as described in the Note below.

---

*If your robot uses four wheel drive, you must configure the **RobotMo tor** library code by modifying the **RobotMotor.h** file to tell the compiler that the library should be built for the 4WD chassis. See **"Installing Third-Party Libraries" (page 83)** for details on how to do this.*

A third library, named IrRemote, is also included in the download. This library won't be needed until Chapter 11, but copying it into your libraries folder now will save you having to do this later.

Sketches Used in This Chapter

- helloRobot.ino—A sketch that rotates the robot when triggered by a sensor. The code uses constants to refer to sensors and motors, and contains functions for handling the infrared reflectance sensors. The sketch uses the RobotMotor library to interface with the motors as shown in Figure 6-4.

- myRobot.ino—The functionality from *helloRobot.ino* restructured into modules using Arduino tabs. Program constants are moved into a tab named robotDefines.h. Reflectance sensor code is moved into a tab named IrSensors as shown in Figure 6-5.

Figure 6-4. *HelloRobot Sketch*

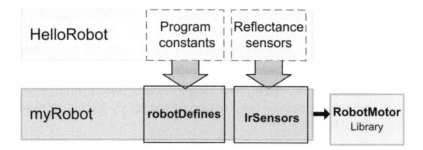

Figure 6-5. *myRobot Sketch*

Load and Run helloRobot.ino

Example 6-1 shows the sketch you can use to test edge detection. Before you upload the sketch, ensure the battery power switch is off (switch toggle angled toward the DC jack) and connect your Arduino to your computer with a USB cable. Next, upload the sketch (see Chapter 5 if you need help loading the sketch).

Example 6-1. **The Hello, Robot sketch**

```
/****************************************************
HelloRobot.ino: Initial Robot test sketch

Michael Margolis 4 July 2012
****************************************************/
// include motor libraries
#include <AFMotor.h>     // adafruit motor shield library
#include <RobotMotor.h>  // 2wd or 4wd motor library

/***** Global Defines ****/
// defines to identify sensors
const int SENSE_IR_LEFT   = 0;
const int SENSE_IR_RIGHT  = 1;
const int SENSE_IR_CENTER = 2;

// defines for directions
const int DIR_LEFT   = 0;
const int DIR_RIGHT  = 1;
const int DIR_CENTER = 2;

const char* locationString[] = {"Left", "Right",    "Center"}; // Debug labels
// http://arduino.cc/en/Reference/String for more on character string arrays

// obstacles constants
const int OBST_NONE       = 0;  // no obstacle detected
const int OBST_LEFT_EDGE  = 1;  // left edge detected
const int OBST_RIGHT_EDGE = 2;  // right edge detected
const int OBST_FRONT_EDGE = 3;  // edge detect at both left and right sensors
```

```
const int LED_PIN = 13;
/**** End of Global Defines ****************/


// Setup runs at startup and is used configure pins and init system variables
void setup()
{
  Serial.begin(9600);
  blinkNumber(8); // open port while flashing. Needed for Leonardo only

  motorBegin(MOTOR_LEFT);
  motorBegin(MOTOR_RIGHT);

  irSensorBegin();     // initialize sensors
  pinMode(LED_PIN, OUTPUT); // enable the LED pin for output

  Serial.println("Waiting for a sensor to detect blocked reflection");
}


void loop()
{
    // call a function when reflection blocked on left side
    if(lookForObstacle(OBST_LEFT_EDGE) == true)   {
      calibrateRotationRate(DIR_LEFT,360);   // calibrate CCW rotation
    }
    // as above for right sensor
    if(lookForObstacle(OBST_RIGHT_EDGE) == true)   {
      calibrateRotationRate(DIR_RIGHT, 360);   // calibrate CW rotation
    }
}


// function to indicate numbers by flashing the built-in LED
void blinkNumber( byte number) {
    pinMode(LED_PIN, OUTPUT); // enable the LED pin for output
    while(number--) {
      digitalWrite(LED_PIN, HIGH); delay(100);
      digitalWrite(LED_PIN, LOW);  delay(400);
    }
}


/**********************
 code to look for obstacles
**********************/

// returns true if the given obstacle is detected
boolean lookForObstacle(int obstacle)
{
  switch(obstacle) {
    case  OBST_FRONT_EDGE: return irEdgeDetect(DIR_LEFT) || irEdgeDetect(DIR_RIGHT);
    case  OBST_LEFT_EDGE:  return irEdgeDetect(DIR_LEFT);
    case  OBST_RIGHT_EDGE: return irEdgeDetect(DIR_RIGHT);
  }
  return false;
}
```

```
/***********************************
 functions to rotate the robot
***********************************/

// return the time in milliseconds to turn the given angle at the given speed
long rotationAngleToTime( int angle, int speed)
{
int fullRotationTime; // time to rotate 360 degrees at given speed

  if(speed < MIN_SPEED)
    return 0; // ignore speeds slower then the first table entry

  angle = abs(angle);

  if(speed >= 100)
    fullRotationTime = rotationTime[NBR_SPEEDS-1]; // the last entry is 100%
  else
  {
    int index = (speed - MIN_SPEED) / SPEED_TABLE_INTERVAL; // index into speed
                                                  // and time tables
    int t0 =  rotationTime[index];
    int t1 = rotationTime[index+1];    // time of the next higher speed
    fullRotationTime = map(speed,
                      speedTable[index],
                      speedTable[index+1], t0, t1);
    // Serial.print("index= ");  Serial.print(index); Serial.print(", t0 = ");
    // Serial.print(t0);  Serial.print(", t1 = ");  Serial.print(t1);
  }
  // Serial.print(" full rotation time = ");  Serial.println(fullRotationTime);
  long result = map(angle, 0,360, 0, fullRotationTime);
  return result;
}

// rotate the robot from MIN_SPEED to 100% increasing by SPEED_TABLE_INTERVAL
void calibrateRotationRate(int sensor, int angle)
{
  Serial.print(locationString[sensor]);
  Serial.println(" calibration" );
  for(int speed = MIN_SPEED; speed <= 100; speed += SPEED_TABLE_INTERVAL)
  {

    delay(1000);
    blinkNumber(speed/10);

    if( sensor == DIR_LEFT)
    {    // rotate left
      motorReverse(MOTOR_LEFT,  speed);
      motorForward(MOTOR_RIGHT, speed);
    }
    else if( sensor == DIR_RIGHT)
    {    // rotate right
      motorForward(MOTOR_LEFT,  speed);
      motorReverse(MOTOR_RIGHT, speed);
```

```
      }
      else
         Serial.println("Invalid sensor");

      int time = rotationAngleToTime(angle, speed);

      Serial.print(locationString[sensor]); Serial.print(": rotate ");
      Serial.print(angle); Serial.print(" degrees at speed "); Serial.print(speed);
      Serial.print(" for "); Serial.print(time); Serial.println("ms");

      delay(time);
      motorStop(MOTOR_LEFT);
      motorStop(MOTOR_RIGHT);
      delay(2000); // two second delay between speeds
   }
}


/****************************
   ir reflectance sensor code
****************************/

const byte NBR_SENSORS = 3;  // this version only has left and right sensors
const byte IR_SENSOR[NBR_SENSORS] = {0, 1, 2}; // analog pins for sensors

int irSensorAmbient[NBR_SENSORS]; // sensor value with no reflection
int irSensorReflect[NBR_SENSORS]; // value considered detecting an object
int irSensorEdge[NBR_SENSORS];    // value considered detecting an edge
boolean isDetected[NBR_SENSORS] = {false,false}; // set true if object detected

const int irReflectThreshold = 10; // % level below ambient to trigger reflection
const int irEdgeThreshold    = 90; // % level above ambient to trigger edge

void irSensorBegin()
{
  for(int sensor = 0; sensor < NBR_SENSORS; sensor++)
      irSensorCalibrate(sensor);
}

// calibrate thresholds for ambient light
void irSensorCalibrate(byte sensor)
{
  int ambient = analogRead(IR_SENSOR[sensor]); // get ambient level
  irSensorAmbient[sensor] = ambient;
  // precalculate the levels for object and edge detection
  irSensorReflect[sensor] = (ambient * (long)(100-irReflectThreshold)) / 100;
  irSensorEdge[sensor]    = (ambient * (long)(100+irEdgeThreshold)) / 100;
}

// returns true if an object reflection detected on the given sensor
// the sensor parameter is the index into the sensor array
boolean irSensorDetect(int sensor)
{
  boolean result = false; // default value
  int value = analogRead(IR_SENSOR[sensor]); // get IR light level
```

```
   if( value <= irSensorReflect[sensor]) {
      result = true; // object detected (lower value means more reflection)
      if( isDetected[sensor] == false) { // only print on initial detection
         Serial.print(locationString[sensor]);
         Serial.println(" object detected");
      }
   }
   isDetected[sensor] = result;
   return result;
}

boolean irEdgeDetect(int sensor)
{
   boolean result = false; // default value
   int value = analogRead(IR_SENSOR[sensor]); // get IR light level
   if( value >= irSensorEdge[sensor]) {
      result = true; // edge detected (higher value means less reflection)
      if( isDetected[sensor] == false) { // only print on initial detection
         Serial.print(locationString[sensor]);
         Serial.println(" edge detected");
      }
   }
   isDetected[sensor] = result;
   return result;
}
```

The sketch tests the calibration of the robot's speed of movement. The front sensors are used to initiate a motor test—the motors rotate the robot 360 degrees in the direction of the sensor that was triggered. If the robot is functioning correctly, it will execute a complete revolution at seven speeds ranging from slowest to fastest.

To run the test, place the robot on a reflective white surface such as a large sheet of paper. When the robot's up and running, Arduino's pin 13 LED will flash once.

Another way to test is to put the robot on something that will raise the wheels off the ground by an inch or so. This will enable the motors to turn without the robot skittering around.

This sketch displays debugging information to the serial console. If you'd like to view it, you'll need to keep the USB cable plugged into your computer and your robot; be careful, since the robot will be moving. If you're using an Arduino

Leonardo, wait until the robot's LED flashes to indicate it's ready before open-ing the Arduino Serial Monitor (the Serial Monitor is the rightmost icon on the Arduino toolbar). When the sketch starts, you should see the following in the Arduino Serial Monitor:

```
Waiting for a sensor to detect blocked reflection
```

Swipe something dark (a small piece of matte black paper the size of a business card works well) near one of the sensors (panel 2 seen in Figure 6-7). The Serial monitor should now display the output similar to that shown in Example 6-2. The number of lines and the values displayed will vary with different robots but you should see multiple lines showing the direction of rotation, speed and time in milliseconds.

Example 6-2. Serial output from HelloRobot.ino

```
Left calibration
    Left: rotate 360 degrees at speed 40 for 5500ms
    Left: rotate 360 degrees at speed 50 for 3300ms
    Left: rotate 360 degrees at speed 60 for 2400ms
    Left: rotate 360 degrees at speed 70 for 2000ms
    Left: rotate 360 degrees at speed 80 for 1750ms
    Left: rotate 360 degrees at speed 90 for 1550ms
    Left: rotate 360 degrees at speed 100 for 1150ms
```

Motors on the left side should spin in reverse, motors on the right should spin forward for the indicated time in milliseconds (if the robot was on the ground, it would rotate to the left (counter-clockwise). If you don't see the expected results, see "Troubleshooting" (page 98) for help.

Completing this test will verify that everything (the robot motors, power source, Arduino and motor shield) is wired up and functioning correctly. Dou-ble check that you have completed all the building steps. Take particular care that the battery wires to the motor shield are attached to the correct polarity.

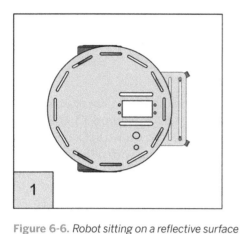

Figure 6-6 shows the robot stationary on a reflective surface. If the robot moves when placed on the surface, switch the power off and then on so the robot can measure and calibrate for the ambient light level. It should remain motionless until a sensor detects a reduction in the light reflected off the surface.

Figure 6-6. Robot sitting on a reflective surface

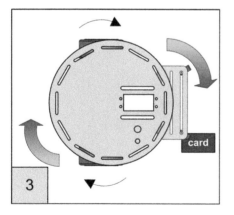

The robot should rotate in the direction of the sensor that detects the reduced reflection. In Figure 6-7, a non-reflective card is swiped under the right sensor which will trigger the robot to turn clockwise.

Figure 6-7. Non-reflective card under right sensor

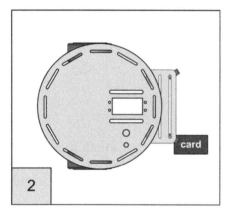

Figure 6-8 shows the right motor running backwards and the left forwards which will rotate the robot clockwise.

Figure 6-8. Robot rotates in direction of swiped sensor

About the Sketch

The code in Example 6-1 forms the nucleus of all the sketches that follow so it is worthwhile taking a moment to look through the code to see how it works. The purpose of the code is to drive the motors in opposite directions for a duration that will rotate the robot one revolution at speeds ranging from the slowest speed to the fastest. All the sketches in the book refer to speeds as a percent of maximum speed, the next chapter, Chapter 7, *Controlling Speed and Direction* explains speed control in detail.

The functions to access the Adafruit motor shield (see Chapter 2, *Building the Electronics*) are included by the line shown in Example 6-3.

Example 6-3. #include line for Adafruit library
```
#include <AFMotor.h>     // adafruit motor shield library
```

The line shown in Example 6-4 includes the library written for this book (RobotMotor.h).

Example 6-4. #include line for this book's library
```
#include <RobotMotor.h>  // 2wd or 4wd motor library
```

This library provides a consistent interface for motor functions in order to isolate the higher level logic from hardware specifics. This means that you can use the same sketch code with (almost) any motor hardware simply by changing the RobotMotor library code to suit the hardware. The motor code is explained in the next chapter and you can find example code to support a different motor controller in Appendix B, *Using Other Hardware with Your Robot*.

The block that begins with: /***** Global Defines ****/ contains declarations for constants that identify: sensors, directions and obstacles. These constants enable you to refer to elements in your sketch using meaningful names instead of numbers, for example this:
```
calibrateRotationRate(DIR_LEFT,360);
```
instead of this:
```
motorForward(0, 360);
```

The setup section calls functions to initialize the motor and sensor modules (more on these later). The loop function uses the lookForObstacle function to determine if a reflection is detected. It waits until no reflection is detected on either sensor; the robot is not on the ground (or on a non-reflective surface). The `lookForObstacle` function is checked to determine if the left or right sensor detects a reflection, and if so, calls the `calibrateRotationRate` function to rotate the robot for a short period.

The `lookForObstacle` function is told which obstacle to check for (the obstacles are identified using the defines described above). The case statement (see *http://arduino.cc/en/Reference/SwitchCase*) is used to call `irEdgeDetect` function that returns true if an object is detected on that sensor. If no object is detected, the function returns `OBST_NONE`, shown in Example 6-5. See "Infrared Reflectance Sensors" (page 134) for a detailed explanation of `irEdgeDetect` and related functions.

Example 6-5. **The lookForObstacle function**

```
// returns true if the given obstacle is detected
boolean lookForObstacle(int obstacle)
{
  switch(obstacle) {
    case  OBST_FRONT_EDGE: return irEdgeDetect(DIR_LEFT) &&
                                  irEdgeDetect(DIR_RIGHT);
    case  OBST_LEFT_EDGE:  return irEdgeDetect(DIR_LEFT);
    case  OBST_RIGHT_EDGE: return irEdgeDetect(DIR_RIGHT);
  }
  return false;
}
```

The sensor detection is done in the function `irSensorDetect`, shown in Example 6-6.

Example 6-6. **The irSensorDetect function**

```
// returns true if reflection level reduces below a threshold
// for example if the robot is does not sense the reflective surface
// the sensor parameter is the index into the sensor array
boolean irEdgeDetect(int sensor)
{
  boolean result = false; // default value
  int value = analogRead(IR_SENSOR[sensor]); // get IR light level
  if( value >= irSensorEdge[sensor]) {
      result = true; // edge detected (higher value means less reflection)
      if( isDetected[sensor] == false) { // only print on initial detection
        Serial.print(locationString[sensor]);
        Serial.println(" edge detected");
      }
```

```
  }
  isDetected[sensor] = result;
  return result;
}
```

This function will return `true` if the reflection level is reduced. This is determined through a call to `analogRead` to get a raw sensor reading that is compared to a detected/not detected threshold. Values greater than or equal to the threshold are considered a loss of reflection (the voltage from the sensor increases when the reflected light decreases , see "Infrared Reflectance Sensors" (page 134) for details on the reflectance sensors). The results from this test is stored in an array named `isDetected`. The array can be used to recall the sensor state of the most recent call to `irSensorDetect` and is used here to suppress printing of the test result if a previous test already indicated that an object was detected, as shown in Example 6-7 .

Example 6-7. **Initial detection**
```
if( isDetected[sensor] == false) { // only print on initial detection
  Serial.print(locationString[sensor]);
  Serial.println(" object detected");
}
```

The motor code commands the motor controller board to drive the motor forwards, backwards or stop.

For example, the following will spin the right motor forward at a speed given by the `speed` parameter (the parameter is the percentage of the maximum speed):
```
motorForward(MOTOR_RIGHT, speed);
```

Motor code is explained in detail in Chapter 7, *Controlling Speed and Direction*.

Rotating the robot is handled by the `calibrateRotationRate` function. For example. if the left sensor is triggered, the code will spin the left motor in reverse and the right motor forward, thus rotating the robot towards the left (counterclockwise):
```
if( sensor == DIR_LEFT)
{  // rotate left
    motorReverse(MOTOR_LEFT,  speed);
    motorForward(MOTOR_RIGHT, speed);
}
```

Troubleshooting

If you are having trouble getting HelloRobot working then the first thing to do is to put the robot down, walk away from your computer screen and have a refreshing drink. Come back and look at things with fresh eyes and check to see if you have things wired up and connected correctly. If it looks like the connections are okay, then the next step is to make a list of the major symptoms:

Compile errors

- `'AF_DCMotor'` does not name a type error message—this message indicates the AFMotor library has not been found. This library is included with the download code for this book (see "How to Contact Us" (page xv) for the URL). See "Installing Third-Party Libraries" (page 83) in Chapter 5, *Tutorial: Getting Started with Arduino* for help with this.

- `"This chip is not supported!"` error message—This message is displayed if the chip selected in the IDE is not recognized by the library. This will occur if you select the Leonardo board and use a version of the AF-motor library that does not support this chip. Replacing your AFMotor library with the one in the book's example code will fix this problem.

- `"expected definition: CHASSIS_2WD or CHASSIS_4WD not found"` will be displayed if you changed the defines in the RobotMotor library to an invalid value. This library expects to find either CHASSIS_2WD or CHASSIS_4WD following the `#define` in RobotMotor.h.

Software Errors

- The Serial Monitor is not displaying the text shown at the end of "Load and Run helloRobot.ino" (page 88)—read through Chapter 5, *Tutorial: Getting Started with Arduino* and check that you have the drivers for your board correctly installed.

- The Serial Monitor displays the initial text but then displays errors or other unexpected text—see Appendix C, *Debugging Your Robot*.

Hardware symptoms

- No LEDs on the Arduino board are lit (you may need to remove the motor shield to check this). - This usually means that either no power is being supplied to the board. If the power switch is on, check that the batteries have sufficient voltage and are located correctly. Check the wiring from the battery and switch to the shield.

- Motors don't turn—Check that the batteries are fitted correctly (USB does not provide enough power to drive the motors). Check the motor wiring.

You can test each motor by disconnecting the motor wires going to the motor terminals on the shield and connecting them directly to the battery terminals. If the motors still do not turn but the shield LED is lit, then double check the shield soldering.

- Two of the four motors don't turn on the 4WD robot—Have you configured the library for 4WD?—see "Software Prerequisites" (page 86).
- Motors run but the robot does not rotate 360 degrees—the robot rotation does not need to be exact; anything within 20 or 30 degrees is good enough. See Chapter 7 if you do want to adjust the rotation rate.

See Appendix C, *Debugging Your Robot* for more on debugging.

Making the Sketch Easy to Enhance

Although this is the simplest sketch in this book, it performs a number of different tasks: controlling the motors, interfacing with sensors, and rotating the robot in response to object detection. You will be adding much more functionality to the robot in later chapters. To help keep the various functional elements under control, it makes sense to organize the code into modules to keep functionally similar code together and to separate code that is not functionally related.

The HelloRobot sketch naturally divides into three sections: the main logic (the loop and rotation code), sensor interface, and motor control.

Moving the sensor interface and motor control into separate modules makes the code easier to enhance. You can change one of the modules without disturbing the code for the other. And you can easily copy modules into other sketches—the tab code file has the same name as shown in the tab, with the extension .ino. Adding this file to another sketch will automatically create the tab for that sketch the next time the sketch is opened on the IDE.

The Arduino IDE provides tabs as a convenient mechanism for managing modules (see Chapter 5, *Tutorial: Getting Started with Arduino*, "Using Tabs" (page 82)). The following explains how sections of HelloRobot code are moved into two new tabs, one providing an interface for IR sensors, the other an interface for motors. All of the code in later chapters use tabs as containers for functional modules.

The following steps creates a sketch named myRobot derived from HelloRo bot that contains two tabs for sensor and motor functions:

You can download the myRobot sketch from the book's website but you may want to go through these steps yourself to familiarize yourself with the procedure for creating and using tabs in the IDE.

1. Load the HelloRobot sketch and use the IDE file menu to save as 'myRobot'.

2. Create a tab by clicking the tab dropdown and selecting 'New Tab' (see Figure 5-8). Name the tab 'IrSensors'.

3. Click the myRobot tab, scroll down to the end of the sketch and cut all code from the end up to the `ir reflectance sensor` code (Example 6-8) comment and paste it into the `IrSensors` tab.

Example 6-8. **IR reflectance sensor code**

```
/****************************
    ir reflectance sensor code
****************************/

const byte NBR_SENSORS = 3;  // this version only has left and right sensors
const byte IR_SENSOR[NBR_SENSORS] = {0, 1, 2}; // analog pins for sensors

int irSensorAmbient[NBR_SENSORS]; // sensor value with no reflection
int irSensorReflect[NBR_SENSORS]; // value considered detecting an object
int irSensorEdge[NBR_SENSORS];    // value considered detecting an edge
boolean isDetected[NBR_SENSORS] = {false,false}; // set true if object detected

const int irReflectThreshold = 10; // % level below ambient to trigger reflection
const int irEdgeThreshold    = 90; // % level above ambient to trigger edge

void irSensorBegin()
{
  for(int sensor = 0; sensor < NBR_SENSORS; sensor++)
     irSensorCalibrate(sensor);
}

// calibrate for ambient light
void irSensorCalibrate(byte sensor)
{
   int ambient = analogRead(IR_SENSOR[sensor]); // get ambient level
  irSensorAmbient[sensor] = ambient;
  // precalculate the levels for object and edge detection
  irSensorReflect[sensor] = (ambient * (long)(100-irReflectThreshold)) / 100;
  irSensorEdge[sensor]    = (ambient * (long)(100+irEdgeThreshold)) / 100;
}

// returns true if an object reflection detected on the given sensor
// the sensor parameter is the index into the sensor array
boolean irSensorDetect(int sensor)
```

```
{
  boolean result = false; // default value
  int value = analogRead(IR_SENSOR[sensor]); // get IR light level
  if( value <= irSensorReflect[sensor]) {
      result = true; // object detected (lower value means more reflection)
      if( isDetected[sensor] == false) { // only print on initial detection
        Serial.print(locationString[sensor]);
        Serial.println(" object detected");
      }
  }
  isDetected[sensor] = result;
  return result;
}

boolean irEdgeDetect(int sensor)
{
  boolean result = false; // default value
  int value = analogRead(IR_SENSOR[sensor]); // get IR light level
  if( value >= irSensorEdge[sensor]) {
      result = true; // edge detected (higher value means less reflection)
      if( isDetected[sensor] == false) { // only print on initial detection
        Serial.print(locationString[sensor]);
        Serial.println(" edge detected");
      }
  }
  isDetected[sensor] = result;
  return result;
}
```

The myRobotOk example sketch provided in the book download code shows the code after the code is moved into the tabs.

Global Definitions

Definitions that need to be accessed across multiple modules are called 'global' definitions. These are generally stored in files called 'header files' (or 'headers'). These files typically have a file extension of .h and the file containing these global definitions is here called robotDefines.h. Although the Arduino build process will automatically make all of the functions in each tab accessible throughout the sketch, constant definitions should be explicitly included at the top of the main tab as follows:

```
// include the global defines
    #include "robotDefines.h"
```

The final step in restructuring the sketch is to move the constant definitions at the top of the sketch into a separate tab. These constants are used by a number of different modules and collecting these together makes it easier to ensure that the values are accessible by all the modules:

1. Create a tab named robotDefines.h (don't forget the .h).

2. From the top of the myRobot tab, move the defines starting from:

```
/**** Global Defines ****/
```

and ending at:

```
/*** End of Global Defines *******/
```

(Example 6-9) into the tab you just created.

3. Switch back to the myRobot tab, and add this line at the top, right after the #includes for *AFMotor.h* and *RobotMotor.h*:

```
#include "robotDefines.h"
```

This is the code that goes into the robotDefines.h:

Example 6-9. Global defines

```
/***** Global Defines ****/

// defines to identify sensors
const int SENSE_IR_LEFT  = 0;
const int SENSE_IR_RIGHT = 1;

// defines for directions
const int DIR_LEFT   = 0;
const int DIR_RIGHT  = 1;
const int DIR_CENTER = 2;

const char* locationString[] = {"Left", "Right",   "Center"}; // Debug labels
// http://arduino.cc/en/Reference/String for more on character string arrays

// obstacles constants
const int OBST_NONE       = 0;  // no obstacle detected
const int OBST_LEFT_EDGE  = 1;  // left edge detected
const int OBST_RIGHT_EDGE = 2;  // right edge detected
const int OBST_FRONT_EDGE = 3;  // edge detect at both left and right sensors

const int LED_PIN = 13;

/**** End of Global Defines ****************/
```

Controlling Speed and Direction

7

This chapter covers the principles of robot motor control that apply to both two wheeled and four wheeled platforms. The motor controller hardware is explained, as is the code used to make this functionality accessible to the sketches. The second half of this chapter ("Software Architecture for Robot Mobility" (page 119)) describes software modules that frees the sketch logic from a dependency on any specific motor hardware. All sketches use the library named RobotMotor that provides a consistent interface to the hardware specific motor system. An optional software module named Move provides high level functions to move the robot that simplifies the code in the more complex sketches that follow in chapters to come.

Hardware Required

- This chapter uses the AFMotor shield described in Chapter 2.

Sketches Used in This Chapter

- The motor control code used in Chapter 6 is explained and two new sketches are introduced:
- MyRobotCalibrateRotation.ino—A sketch for running the robot through a range of speeds to calibrate the robot.

- `MyRobotMove.ino`—This sketch shows how to use higher level movement functions. Constants for defining the current robot movement are added to the robotDefines tab. A new tab named Move is added that contains the high level movement functions. The IrSensor tab and RobotMotor library are unchanged (Figure 7-1).

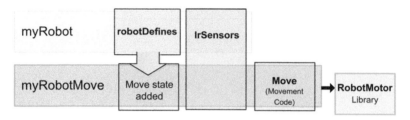

Figure 7-1. *myRobotMove Sketch*

Types of Motors

Brushed DC Motors, such as the ones used in the two wheeled and four wheeled platforms (see Figure 7-2) are the most common type used with Arduino robots. These have two leads connected to brushes (contacts) that control the magnetic field of the coils that drive the motor core (armature). Motor direction can be reversed by reversing the polarity of the power source. These motors typically rotate too fast to directly drive the robot wheels or tracks, so gear reduction is used to reduce speed and increase torque.

Figure 7-2. *DC motor with gearbox*

Other kinds of motors can be used to power robots; here are some you may come across:

Continuous rotation servo

These motors are used on smaller robots. They have the advantage that the motor controller, motor, and gearbox are all mounted in the same housing, so they are easy to attach to a robot and can be driven directly from Arduino pins. However they usually have less torque than typical stand-alone brushed motors.

Brushless motors

These have increased torque and efficiency compared to brushed motors but they are more expensive and complex to control. However, prices are dropping and they are a good choice for a larger robot.

Stepper motors

These motors are used on large robots when precise control is required. These motors typically require 12 or 24 volts so they are not often used on small battery operated robots. However they may become more popular due to the recent availability of low cost 5 volt steppers.

Motor Controllers

The two wheel and four wheel platforms use small DC motors that are controlled using an H-Bridge. The H-Bridge featured in this book is part of the AFMotor shield from Adafruit Industries. This can drive up to four motors independently, although only two are used with the two wheeled robot. This shield requires a library for interfacing sketch code with the hardware; this library is included with the code download for this book (see "How to Contact Us" (page xv)).

The name H-bridge derives from the characteristic shape that you can see in these figures.

To enable the sketches to work with other H-Bridge hardware, a library named RobotMotor is provided with the example code that provides generic control functions that the library translates into the specific commands for the AFMotor shield or another shield if you have use different hardware. see "Software Architecture for Robot Mobility" (page 119)

*This library is modified from the one on the Adafruit site to work with the Leonardo board. The standard Adafruit library can be used with the Uno board). See **"Installing Third-Party Libraries" (page 83)** if you need directions for installing a library. If you followed along with **Chapter 6**, you will already have the library installed.*

The following diagrams explain how an H-bridge works and the `RobotMotor` functions used to control the motors:

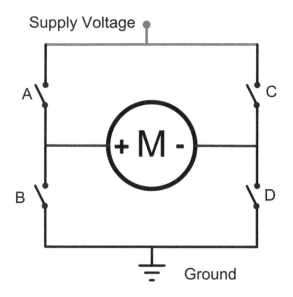

Supply Voltage

A

C

+ M -

B

D

Ground

Figure 7-3. H-Bridge with Motor Idle

Figure 7-3 is a schematic drawing that shows how an H-bridge works. The motor is connected to the positive supply voltage and ground through four switches (in the actual H-bridge, the switching is done with transistors). When all the switches are open, no current flows and the motor is stopped. The code to stop a motor is:

```
motorStop(motor);
```

The parameter in brackets (`motor`) is a constant identifying the motor (`MOTOR_LEFT` or `MOTOR_RIGHT`) to control. The software for the four wheeled robot treats the two motors on the same side as if they were a single motor.

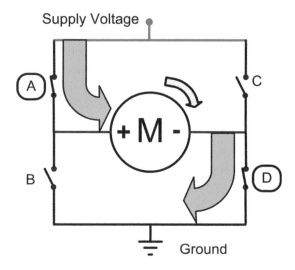

Supply Voltage

A

C

+ M -

B

D

Ground

Figure 7-4. H-Bridge with Motor Running Forward

Figure 7-4 shows the two switches that when closed will cause the motor to run forward. The one marked A connects the positive motor terminal to the positive power supply. Switch D connects the negative motor terminal to ground. The code to run a motor forward at the given speed is:

```
motorForward(motor, speed);
```

The constant speed is a value representing speed as a percent of maximum speed. This is described in the section: "Controlling Motor Speed" (page 109).

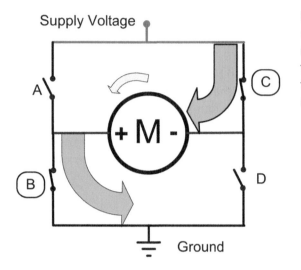

Figure 7-5. H-Bridge with Motor Running in Reverse

Figure 7-5 shows that the opposite switches result in the motor reversing. Switch C connects the positive supply to the negative motor terminal. Switch B connects the positive motor terminal to ground. The code to run a motor in reverse at the given speed is:

```
motorReverse(motor, speed);
```

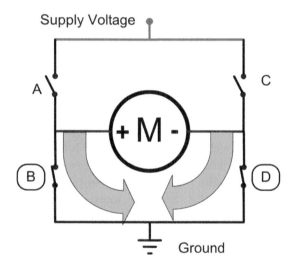

Figure 7-6. H-Bridge with Motor Brake

Figure 7-6 shows switches B and D closed. Both motor terminals are connected together - neither terminal is connected to the positive supply voltage. This is a mode supported by some H-bridge hardware to stop the motor more quickly than it would in the previous case where the motor is simply disconnected from the power. Because the motor terminals are shorted, the motor will resist rotation. If the motor had been spinning and then set to this mode, it will stop more quickly than if the terminals were simply disconnected. The code to brake a motor is:

```
motorBrake(motor);
```

Not all H-bridges, including the Adafruit library, support this mode. The RobotMotor library will call motorStop when the motorBrake function is called.

Controlling Motor Speed

How Motor Speed Is Controlled

Motor speed is controlled by a technique called *Pulse Width Modulation* (PWM), which varies the proportion of the motors on-time to-off time. The higher the proportion of on-time, the greater the motor power and the faster the robot will move (see Figure 7-7).

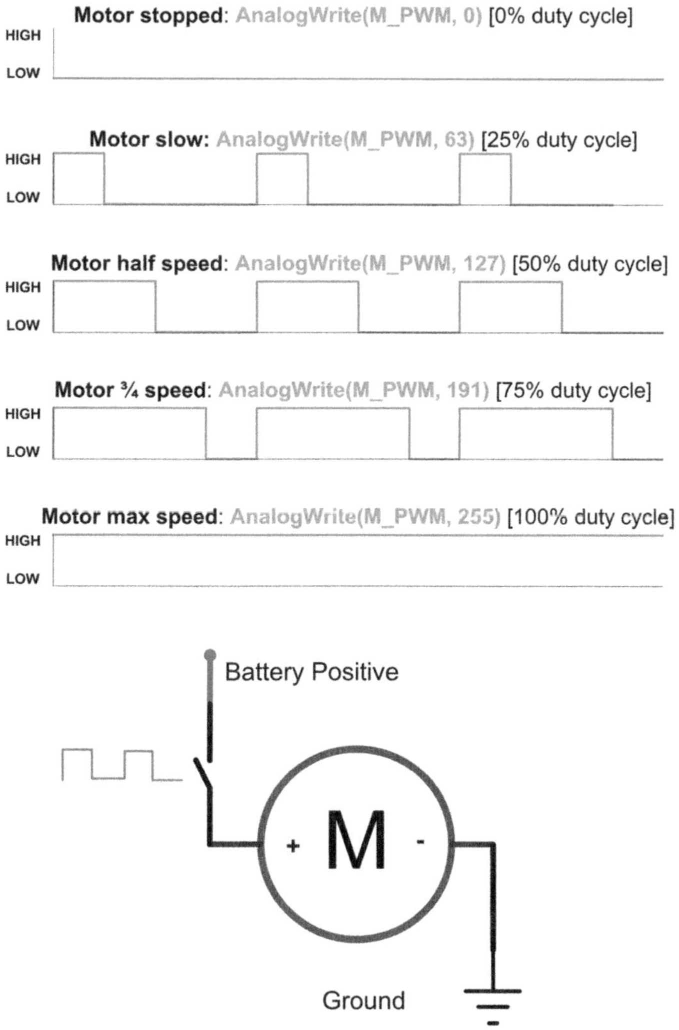

Motor stopped: AnalogWrite(M_PWM, 0) [0% duty cycle]

HIGH

LOW

Motor slow: AnalogWrite(M_PWM, 63) [25% duty cycle]

HIGH

LOW

Motor half speed: AnalogWrite(M_PWM, 127) [50% duty cycle]

HIGH

LOW

Motor ¾ speed: AnalogWrite(M_PWM, 191) [75% duty cycle]

HIGH

LOW

Motor max speed: AnalogWrite(M_PWM, 255) [100% duty cycle]

HIGH

LOW

Battery Positive

+ M -

Ground

Figure 7-7. *Controlling motor power using Pulse Width Modulation*

All the code in this book uses a percent value to refer to speed. This value is the percentage of power given to the motors (technically, its called the *duty cycle*). Percent speed is used instead of the raw PWM value to isolate the sketch logic from the low level motor code. Different motor hardware use various techniques for controlling motor rotation speed. For example, continuous rotation servos can use servo angle (where 90 is stop and 0 actually rotates the motor at full reverse speed), and stepper motors don't use PWM to control speed. Because the high level logic always uses percent and this is mapped to the range needed by the hardware in the motor interface code, the same high level code can be used with other hardware simply by swapping the appropriate motor interface module.

Code for Motor Control

The AFMotor library that interfaces with the motor hardware expects a PWM value ranging from 0 to 255, so the motorSetSpeed function in the RobotMotor library converts the percent into a PWM value using the map function, as shown in Example 7-1.

Example 7-1. **Setting the motor speed; from RobotMotor.cpp**

```
void motorSetSpeed(int motor, int speed)
{
  motorSpeed[motor] = speed;          // save the value
  int pwm = map(speed, 0,100, 0,255); // scale to PWM range

  motors[motor].setSpeed(pwm) ;
}
```

map is a handy function that is used extensively throughout this book. The function scales a value from one range to another range. For example, the following scales a value from analogRead (0-1023) to a percent (0-100):

```
int toPercent = map(val, 0,1023, 0,100);
```

You can read more about map here: ***http://arduino.cc/en/Reference/map***.

Bear in mind that the speed percentage is actually controlling motor power and this is usually not directly proportional to speed, particularly at low power.

The amount of power required to get the robot moving is dependent on the motor, gearbox, battery voltage, robot weight and the surface the robot is on. The method to calibrate the robot will be described shortly, but first, here is an explanation of how the software handles robot speed control.

The code fragment shown in Example 7-2 contains the constants that are used to calculate the appropriate delays for different speeds to rotate the robot. rotationTime stores the duration for a 360 degree rotation for all practical speeds. Speeds less than MIN_SPEED (40%) do not provide sufficient power to overcome friction in the drive system.

Example 7-2. **Constants for the delays needed to rotate the robot, from RobotMotor.cpp**

```
const int MIN_SPEED = 40; // first table entry is 40% speed
const int SPEED_TABLE_INTERVAL = 10; // each table entry is 10% faster speed
const int NBR_SPEEDS =  1 + (100 - MIN_SPEED)/ SPEED_TABLE_INTERVAL;

int speedTable[NBR_SPEEDS]   = {40,     50,   60,   70,   80,   90,  100}; // speeds
int rotationTime[NBR_SPEEDS] = {5500, 3300, 2400, 2000, 1750, 1550, 1150}; // time
```

The table holds durations in milliseconds for speeds in intervals of 10%. The values were derived from experimentation with the two wheeled robot using a sketch named myRobotCalibrateRotation sketch and noting the angles for each of the speeds as shown in Figure 7-8.

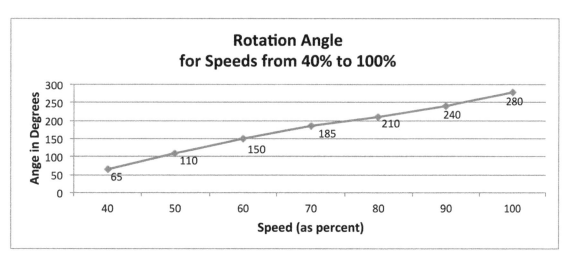

Figure 7-8. *Angle that the robot rotates for one second burst at each of the supported speeds*

By calculating the angle as a fraction of 360 degrees, the time to rotate the robot one complete revolution can be determined for each speed (the calculation for the value in milliseconds is: 1000*(360/angle)).

Figure 7-9 shows the actual times for the 2WD robot.

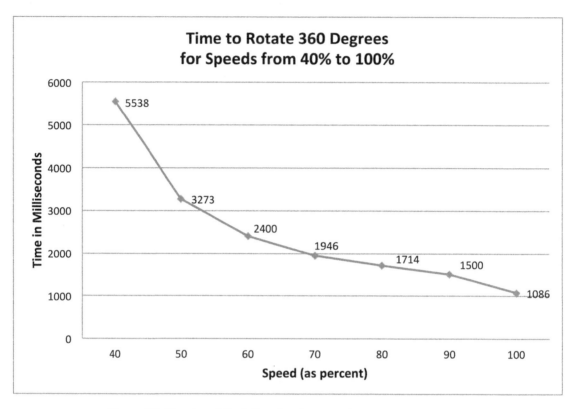

Figure 7-9. *Time for a full rotation at various speeds*

The relationship between rotation angle and speed percentage is not linear, so interpolation is used to calculate the duration to produce a full rotation for any speed (as long as it is as fast or faster than the minimum speed).

Example 7-3 shows the code that uses the table with times based on the data shown in Figure 7-9.

The RobotMotor library has the code to determine how much time the robot requires to rotate 360 degrees. This will differ between the two and four wheeled chassis and vary as the motor speed varies. Example 7-3 shows the values used in the RobotMotor.cpp code for the 2WD chassis.

Example 7-3. **Controlling rotation rate**

```
// tables hold time in ms to rotate robot 360 degrees at various speeds
// this enables conversion of rotation angle into timed motor movement
// The speeds are percent of max speed
// Note: low cost motors do not have enough torque at low speeds so
// the robot will not move below this value
// Interpolation is used to get a time for any speed from MIN_SPEED to 100%

const int MIN_SPEED = 40; // first table entry is 40% speed
const int SPEED_TABLE_INTERVAL = 10; // each table entry is 10% faster speed
const int NBR_SPEEDS =  1 + (100 - MIN_SPEED)/ SPEED_TABLE_INTERVAL;

int speedTable[NBR_SPEEDS]   = {40,    50,   60,   70,   80,   90,  100}; // speeds
int rotationTime[NBR_SPEEDS] = {5500, 3300, 2400, 2000, 1750, 1550, 1150}; // time
```

Example 7-4 shows the values for the 4WD chassis.

Example 7-4. **Controlling rotation rate**

```
const int MIN_SPEED = 60; // first table entry is 60% speed
const int SPEED_TABLE_INTERVAL = 10; // each table entry is 10% faster speed
const int NBR_SPEEDS =  1 + (100 - MIN_SPEED)/ SPEED_TABLE_INTERVAL;

int speedTable[NBR_SPEEDS]   = {60,   70,   80,   90,  100}; // speeds
int rotationTime[NBR_SPEEDS] = {5500, 3300, 2400, 2000, 1750}; // time
```

Note that there are fewer entries in the tables for the 4WD robot because this chassis requires a higher speed to get going. "Calibrating Rotation and Tracking" (page 116) explains how to adjust the tables to suit your robot.

The table entries assume speed intervals of 10% so the value for MIN_SPEED should be multiple of 10. There must be one rotation time per speed so if you increase MIN_SPEED by 10 for example, you will also need to remove the first element in both speedTable and rotationTime.

The code in *RobotMotor.cpp* that uses the data in the rotationTime table is the same for both chassis (see Example 7-5).

Modifying a Library

You know how to modify an Arduino sketch—just edit it in the Arduino IDE. But modifying a library is a bit more involved. You need to go into the sketch folder, open up the library directory, and find the file. Then you need to open it in a text editor. Here's how to modify the *RobotMotor.h* file to use the 4WD chassis.

First, find the sketchbook location. Go to Arduino's preferences (File→Preferences on Windows or Linux, Arduino→Preferences on Mac). Under Sketchbook Location, you'll find the name of the directory that contains your sketches and libraries. Next:

1. Open the sketchbook folder in the Finder (Mac) or Explorer (Windows).

2. Locate the libraries directory inside, and then open the directory named *RobotMotor*.

3. Right-click (or Control-click on the Mac) the *RobotMotor.h* file, and open it with a plain text editor. On Windows, you should use Notepad. On the Mac, you can use TextEdit. On Linux, use your favorite plain text editor.

4. Change #define CHASSIS_2WD to #define CHASSIS_4WD and save the file.

Although you need to quit and restart the Arduino IDE when you install a new library, you don't need to do so each time you modify a library.

Example 7-5. Applying the rotationTime table

```
// return the time in milliseconds to turn the given angle at the given speed
long rotationAngleToTime( int angle, int speed)
{
int fullRotationTime; // time to rotate 360 degrees at given speed

  if(speed < MIN_SPEED)
    return 0; // ignore speeds slower then the first table entry

  angle = abs(angle);

  if(speed >= 100)
    fullRotationTime = rotationTime[NBR_SPEEDS-1]; // the last entry is 100%
  else
  {
    int index = (speed - MIN_SPEED) / SPEED_TABLE_INTERVAL ; // index into speed
                                                // and time tables
    int t0 =  rotationTime[index];
    int t1 = rotationTime[index+1];     // time of the next higher speed
    fullRotationTime = map(speed,
                        speedTable[index],
                        speedTable[index+1], t0, t1);
    // Serial.print("index= ");  Serial.print(index);
    // Serial.print(", t0 = ");  Serial.print(t0);
    // Serial.print(", t1 = ");  Serial.print(t1);
  }
```

```
// Serial.print(" full rotation time = ");  Serial.println(fullRotationTime);
long result = map(angle, 0,360, 0, fullRotationTime);
return result;
}
```

This code determines the index into the speedTable array that is closest to (but not greater than) the desired speed. This index is stored in the variable t0. The interpolated time will be between this value and the next index (t1), with the rotation time calculated using the ratio of the rotationTime value between t0 and t1 in the same proportion as the desired speed in the speedTable. It may be easier to understand how this works by consulting Figure 7-10.

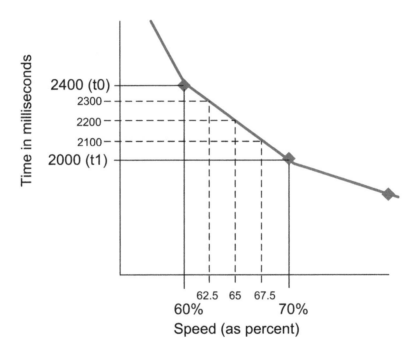

Figure 7-10. *Speed Interpolation*

For example, for a speed of 65%, which is halfway between the values for 60% and 70%, the time associated with 65% speed will be 2200, which is half way between 2400 (the 60% speed value) and 2000 (the 70% speed value). A speed of 62.5% is 1/4 of the range between the table entries (60 and 70), so the time will be 1/4 of the range between the speeds for that range (2400 and 2000, which is 2300 milliseconds). The map function is used to calculate this proportional value:

```
fullRotationTime = map(speed,speedTable[index],speedTable[index+1],t0,t1);
```

To calculate the time to rotate an angle other than 360 degrees, the map function is used again:

```
long result = map(angle, 0,360, 0, fullRotationTime);
```

Calibrating Rotation and Tracking

Motor timings do not need to be exact but if you are using the four wheeled platform you will probably want to calibrate the values in the table because this platform requires more rotation time than the two wheeled version. You can calibrate your robot with the myRobotCalibrateRotation sketch. Here is the main tab for that sketch; the actual calibration is performed in the calibrateSpeed function shown in Example 7-6.

Example 7-6. Robot calibration

```
/********************************************************
MyRobotCalibrateRotation.ino
********************************************************/
// include motor libraries
#include <AFMotor.h>      // adafruit motor shield library
#include <RobotMotor.h>  // 2wd or 4wd motor library

// Setup runs at startup and is used configure pins and init system variables
void setup()
{
  motorBegin(MOTOR_LEFT);
  motorBegin(MOTOR_RIGHT);
  calibrateSpeed();
}

void loop()
{
}

void calibrateSpeed()
{
  for(int speed = MIN_SPEED; speed <= 100; speed += 10)
  {
     // rotate robot left for 1 second
     motorReverse(MOTOR_LEFT,  speed);
```

```
    motorForward(MOTOR_RIGHT, speed);
    delay(1000); // delay 1 second
    motorStop(MOTOR_LEFT);
    motorStop(MOTOR_RIGHT);

    delay(3000);  // wait 3 seconds

    // rotate robot right for 1 second
    motorReverse(MOTOR_RIGHT,  speed);
    motorForward(MOTOR_LEFT, speed);
    delay(1000);  // delay 1 second
    motorStop(MOTOR_LEFT);
    motorStop(MOTOR_RIGHT);
    delay(3000); // wait 3 seconds
  }
}
```

Running this sketch will rotate the robot left (CCW) for one second, stop for one second, then rotate the robot right (CW) for a second. If you mark the angle of the robot after each CCW rotation, you can calculate how much longer or shorter it would take the robot to turn 360 degrees for each speed. If your robot does not rotate at all at the slower speeds, note the lowest speed that the robot does move and set MIN_SPEED in RobotMotor.cpp to this value.

The RobotMotor library also supports the ability to adjust the relative power to each motor in order to prevent the robot drifting off a straight course due to differences in performance between the left and right motor(s). If your robot does not track a straight line when moving forward or backward, you can modify the motor library (see next section) to correct this.

The RobotMotor.cpp library file contains a constant that can be adjusted to correct drift:

```
const int differential = 0; // % faster left motor turns compared to right
```

Here is how the differential constant is used in the code:

```
if( motor == MOTOR_LEFT
    && speed > differential)
    speed -= differential;
```

If your robot drifts, adjust the constant differential to compensate. Set the value using trial and error, positive values nudge the robot to the right, negative values to the left. The correct value will be the difference in speed between the motors in percent. The drift will vary somewhat with motor speed so best to set this when testing with the robot running at a speed midway between the minimum and maximum speeds.

Here is a modified version of the previous sketch that will drive the robot in a straight line when the differential constant is adjusted to correct drift. You can make differential a negative number if your right motor turns faster than your left (the robot drifts to the left).

Example 7-7. **Robot tracking**

```
/**********************************************************
MyRobotCalibrateTracking.ino
**********************************************************/
// include motor libraries
#include <AFMotor.h>      // adafruit motor shield library
#include <RobotMotor.h>   // 2wd or 4wd motor library

const int  TEST_SPEED = MIN_SPEED + 10;    // Typical speed to run the robot
const int differential = 0;   // % faster left motor turns compared to right

// Setup runs at startup and is used configure pins and init system variables
void setup()
{
  motorBegin(MOTOR_LEFT);
  motorBegin(MOTOR_RIGHT);
  calibrateDrift();
}

void loop()
{
}

void calibrateDrift()
{
  motorForward(MOTOR_LEFT, TEST_SPEED - differential);
  motorForward(MOTOR_RIGHT,  TEST_SPEED);
  delay(2000); // delay 2 second
  motorStop(MOTOR_LEFT);
  motorStop(MOTOR_RIGHT);
}
```

If the robot drifts the right when running this sketch, try setting `differential` to 2. If this overcorrects (the robot now drifts to the left), decrease the differential value. If you need more correction, increase the value. If the robot was drifting to the left, use negative values of differential to compensate. You should be able to get the robot running more or less straight after a little trial and error. Don't worry about minor deviations which are caused by small differences in the efficiency of the motors at varying battery levels.

After you have settled in a value for `differential` you must change this in the RobotMotor.cpp file. Open this file with a text editor and (see "Modifying a Library" (page 114)) and find the declaration towards the beginning of the file:

```
const int differential = 0; // % faster left motor turns compared to right
```

Replace 0 with the value determined from the calibration sketch and save the file.

Software Architecture for Robot Mobility

This book provides software modules to minimize the coupling between the application logic and the hardware that actually moves the robot. This minimizes changes that would otherwise be required if you want to use the same logic with different hardware. A low level motor interface library named `RobotMotor` encapsulates the motor hardware functions so that the same sketch code can be used with the two wheeled or four wheeled robot with the Adafruit shield or with a different motor shield. A higher level module named `Move` is also provided to enable the sketch logic to deal with robot movements instead of motor power, for example, the `Move` module has commands to move the robot left or right, to move backward, or rotate 90 degrees. Figure 7-11 shows how the software and hardware is layered. The high level `Move` code is described in detail later in this chapter.

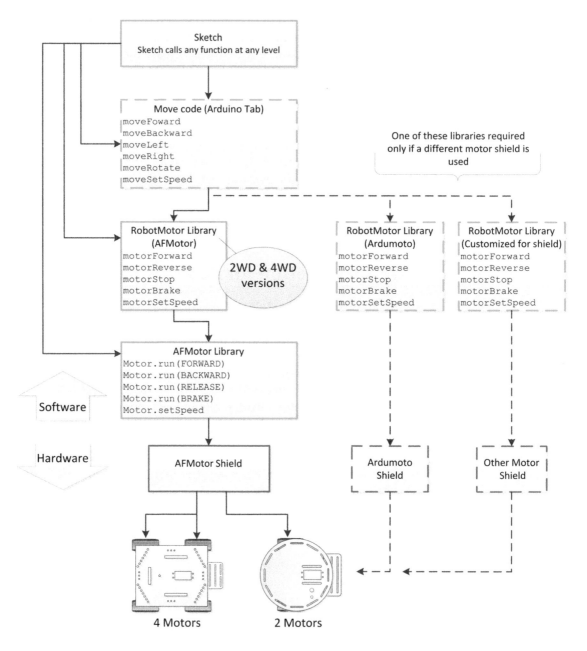

Figure 7-11. *Software architecture for motor control*

Example 7-9 shows the source code for the RobotMotor library's .cpp file. The header file RobotMotor.h (Example 7-8) defines the constants for the left and right motors and declares the functions for speed and direction.

Example 7-8. *RobotMotor.h header file*

```
/*******************************************************
    RobotMotor.h
    low level motor driver interface
    Copyright Michael Margolis May 8 2012
*******************************************************/
/* if you have the 4WD chassis, change the line:
    #define CHASSIS_2WD
  to:
    #define CHASSIS_4WD
 */
#define CHASSIS_2WD // change suffix from 2WD to 4WD if using the 4WD chassis
// defines for left and right motors
const int MOTOR_LEFT  = 0;
const int MOTOR_RIGHT = 1;
extern const int MIN_SPEED;
extern int speedTable[];
extern int rotationTime[];
extern const int SPEED_TABLE_INTERVAL;
extern const int NBR_SPEEDS;
void motorBegin(int motor);
// speed range is 0 to 100 percent
void motorSetSpeed(int motor, int speed);
void motorForward(int motor, int speed);
void motorReverse(int motor, int speed);
void motorStop(int motor);
void motorBrake(int motor);
```

Example 7-9. *RobotMotor functions*

```
/*******************************************************
    RobotMotor.cpp // Adafruit version for 2WD and 4WD chassis
    low level motor driver for use with adafruit motor shield
    Motor constants used are defined AFMotor.h
    Copyright Michael Margolis May 8 2012
*******************************************************/
#include <Arduino.h>
#include <AFMotor.h>  // adafruit motor shield library
#include "RobotMotor.h"
const int differential = 0; // % faster left motor turns compared to right
// tables hold time in ms to rotate robot 360 degrees at various speeds
// this enables conversion of rotation angle into timed motor movement
// The speeds are percent of max speed
// Note: low cost motors do not have enough torque at low speeds so
// the robot will not move below this value
// Interpolation is used to get a time for any speed from MIN_SPEED to 100%
// constants for 2 wheeled robot chassis
#if defined CHASSIS_2WD
const int MIN_SPEED = 40; // first table entry is 40% speed
const int SPEED_TABLE_INTERVAL = 10; // each table entry is 10% faster speed
const int NBR_SPEEDS =  1 + (100 - MIN_SPEED)/ SPEED_TABLE_INTERVAL;
```

```
int speedTable[NBR_SPEEDS]  =  {40,     50,   60,   70,   80,   90,  100}; // speeds
int rotationTime[NBR_SPEEDS] = {5500, 3300, 2400, 2000, 1750, 1550, 1150}; // time
AF_DCMotor motors[] = {
    AF_DCMotor(1, MOTOR12_1KHZ), // left is Motor #1
    AF_DCMotor(2, MOTOR12_1KHZ)  // right is Motor #2 };
// constants for 4 wheeled robot
#elif defined CHASSIS_4WD
const int MIN_SPEED = 60; // first table entry is 60% speed
const int SPEED_TABLE_INTERVAL = 10; // each table entry is 10% faster speed
const int NBR_SPEEDS =  1 + (100 - MIN_SPEED)/ SPEED_TABLE_INTERVAL;

int speedTable[NBR_SPEEDS]  =  {60,    70,   80,   90,  100}; // speeds
int rotationTime[NBR_SPEEDS] = {5500, 3300, 2400, 2000, 1750}; // time
AF_DCMotor motors[] = {
    AF_DCMotor(4, MOTOR34_1KHZ), // left front is Motor #4
    AF_DCMotor(3, MOTOR34_1KHZ), // right front is Motor #3
    AF_DCMotor(1, MOTOR12_1KHZ), // left rear is Motor #1
    AF_DCMotor(2, MOTOR12_1KHZ)  // right rear is Motor #2
};
#else
#error "expected definition: CHASSIS_2WD or CHASSIS_4WD not found"
#endif
int  motorSpeed[2]  = {0,0}; // left and right motor speeds stored here (0-100%)
void motorBegin(int motor)
{
  motorStop(motor);   // stop the front motor
#if defined CHASSIS_4WD
  motorStop(motor+2); // stop the rear motor
#endif
}
// speed range is 0 to 100 percent
void motorSetSpeed(int motor, int speed)
{
  if( motor == MOTOR_LEFT && speed > differential)
    speed -= differential;
  motorSpeed[motor] = speed;              // save the value
  int pwm = map(speed, 0,100, 0,255);   // scale to PWM range

  motors[motor].setSpeed(pwm) ;
#if defined CHASSIS_4WD
  motors[motor+2].setSpeed(pwm) ;
#endif
}
void motorForward(int motor, int speed)
{
  motorSetSpeed(motor, speed);
  motors[motor].run(FORWARD);
#if defined CHASSIS_4WD
  motors[motor+2].run(FORWARD);
#endif
}
void motorReverse(int motor, int speed)
{
```

```
  motorSetSpeed(motor, speed);
  motors[motor].run(BACKWARD);
#if defined CHASSIS_4WD
  motors[motor+2].run(BACKWARD);
#endif
}
void motorStop(int motor)
{
  // todo set speed to 0 ???
  motors[motor].run(RELEASE);      // stopped
#if defined CHASSIS_4WD
  motors[motor+2].run(RELEASE);
#endif
}
void motorBrake(int motor)
{
  motors[motor].run(BRAKE);        // stopped
#if defined CHASSIS_4WD
  motors[motor+2].run(BRAKE);
#endif
}
```

The *RobotMotor.cpp* file contains code for both the two wheel and four wheel chassis. Conditional compilation is used to build the library for the appropriate version. `#if defined CHASSIS_2WD` and `#if defined CHASSIS_4WD` are checks to see which chassis has been defined in the *RobotMotor.h* file. code between `#if defined CHASSIS_2WD` and `#endif` will only be compiled if `CHASSIS_2WD` is defined in *RobotMotor.h*. See "Installing Third-Party Libraries" (page 83) for more details on changing the define for the four wheel chassis.

This library can be modified to support different hardware. For example, see Appendix B for the code to use the Ardumoto shield (but note that Ardumoto only supports two motors so is not suitable for the four wheeled robot).

Functions to Encapsulate Robot Movements

You can simplify your code for controlling your robot's behaviour by using higher level movement functions provided in the Move module. These functions reference the desired movement from the robot's perspective rather than specific motor control. For example, to rotate the robot, rather than calling functions to run one motor forwards and the other backwards, you can call a single function that rotates the robot. And by calibrating the speed of rotation, you can easily get the robot to rotate to any desired angle.

The sketch named myRobotMove has the movement code in a tab called Move. That sketch is similar to the myRobot sketch from "Making the Sketch Easy to

Enhance" (page 99) but uses the rotation functions in the Move tab to drive the robot. Using the higher level functions to drive the robot not only simplifies your code, it isolates the sketch logic from the hardware specific motor code. The sketches in all of the following chapters control robot movement through the functions in the Move tab.

Core Movement Code

Here is a list of the core movement functions:

Move Forward
> Both motors are driven forward at the same speed

Move Backward
> Both motors driven in reverse at the same speed

Move Left
> Left motor stopped, right motor driven forward

Move Right
> Right motor stopped, Left motor driven forward

Move Stop
> Both motors stopped

Set Move Speed
> Used to set the speed for future robot movements

Example 7-10 shows the code in the Move tab that provides the core movement functionality.

Example 7-10. **The core movement functions**

```
/***********************************
 Drive: mid level movement functions
***********************************/

int moveState = MOV_STOP;    // what robot is doing

int   moveSpeed     = 0; // move speed stored here (0-100%)
int   speedIncrement = 10; // percent to increase or decrease speed

void moveBegin()
{
    motorBegin(MOTOR_LEFT);
    motorBegin(MOTOR_RIGHT);
    moveStop();
}

void moveLeft()
{
  motorForward(MOTOR_LEFT,  0);
```

```
    motorForward(MOTOR_RIGHT, moveSpeed);
    changeMoveState(MOV_LEFT);
}

void moveRight()
{
    motorForward(MOTOR_LEFT,  moveSpeed);
    motorForward(MOTOR_RIGHT, 0);
    changeMoveState(MOV_RIGHT);
}

void moveStop()
{
    motorStop(MOTOR_LEFT);
    motorStop(MOTOR_RIGHT);
    changeMoveState(MOV_STOP);
}

void moveBrake()
{
    motorBrake(MOTOR_LEFT);
    motorBrake(MOTOR_RIGHT);
    changeMoveState(MOV_STOP);
}

void moveBackward()
{
    motorReverse(MOTOR_LEFT, moveSpeed);
    motorReverse(MOTOR_RIGHT, moveSpeed);
    changeMoveState(MOV_BACK);
}

void moveForward()
{
    motorForward(MOTOR_LEFT,  moveSpeed);
    motorForward(MOTOR_RIGHT, moveSpeed);
    changeMoveState(MOV_FORWARD);
}

void moveSetSpeed(int speed)
{
    motorSetSpeed(MOTOR_LEFT, speed) ;
    motorSetSpeed(MOTOR_RIGHT, speed) ;
    moveSpeed = speed; // save the value
}
```

The code provides functions that combine the individual motor commands described in "Motor Controllers" (page 106). For example, the moveForward function calls the individual functions to rotate the left and right motors in the direction that moves the robot forward. The speed to move is set by the move SetSpeed function. moveSetSpeed commands the motors to run at the desired speed and stores the speed value so the robot can resume running at the last set speed following an evasive action needed to avoid obstacles.

Additional Core Functions

Some additional functions are included in this tab that are not used in any of the sketches in this book but are convenient if you want to slow down or speed up the robot, for example with remote control. The moveSlower and moveFast er functions can be used to command the robot to decrease or increase speed:

Example 7-11. **Functions to speed up or slow down the robot**

```
void moveSlower(int decrement)
{
   Serial.print(" Slower: ");
   if( moveSpeed >= speedIncrement + MIN_SPEED)
     moveSpeed -= speedIncrement;
   else moveSpeed = MIN_SPEED;
   moveSetSpeed(moveSpeed);
}

void moveFaster(int increment)
{
  Serial.print(" Faster: ");
  moveSpeed += speedIncrement;
  if(moveSpeed > 100)
     moveSpeed = 100;
  moveSetSpeed(moveSpeed);
}

int moveGetState()
{
 return moveState;
}

// this is the low level movement state.
// it will differ from the command state when the robot is avoiding obstacles
void changeMoveState(int newState)
{
  if(newState != moveState)
  {
    Serial.print("Changing move state from "); Serial.print( states[moveState]);
    Serial.print(" to "); Serial.println(states[newState]);
    moveState = newState;
  }
}
```

The moveFaster function increases the current speed by a specified increment and calls moveSetSpeed to make this the current speed. For example, movefaster(10); will result in the robot moving at 85% speed if it was previously moving at 75%.

The moveSlower function is similar but decreases rather than increases the speed. Both functions check to ensure that the new speed is valid. If moveSlower(20) was called when the robot was moving at 85% speed, the robot would slow down to run at 65% speed.

The movement functions also call the function changeMoveState to store the current movement state. These states are defined in the robotDefines.h tab (see Example 6-9) and are used to enable the robot to make decisions with the knowledge of what it is currently doing. For example, detecting an obstacle in front can be handled differently depending on whether the robot is moving forwards or backwards. The robot can check the current move state when it encounters an object and take action if the robot is moving towards it but ignore obstacles that are not in the direction of movement. Here are all the move states:

```
enum {MOV_LEFT, MOV_RIGHT, MOV_FORWARD,
      MOV_BACK, MOV_ROTATE, MOV_STOP};
```

---

*If you are unfamiliar with enum (enumerated lists), see **"Code Style (About the Code)" (page xii)** in **Preface** or an online C or C++ reference.*

To assist debugging, each state has an associated text label that can be printed to the serial monitor to show what the robot should be doing.

```
const char* states[] = {"Left", "Right", "Forward",
                        "Back", "Rotate", "Stop"};
```

The move state defines are located at the end of the robotDefines.h tab.

Functions to Rotate the Robot

Rotation is a common task as the robot is exploring and moving to avoid obstacles. The robots described in this book do not know the angle they are facing or how much actual movement results from driving the motors. Commands to rotate the robot at particular angle are implemented by timing how long to turn the motors based on data collected during calibration. Example 7-12 shows the rotation functions from the Move tab:

Move Rotate
> One motor forward, one reverse for the duration to rotate the robot to the given angle. Positive angles rotate clockwise, negative angles counter-clockwise

Rotation Angle to Time

Function used to calculate the duration to rotate the robot to a given angle at a given speed. This function is the same as listed in "Load and Run helloRobot.ino" (page 88)

Calibrate Rotation Rate

Function used for calibration—the robot will attempt to rotate at a given angle at speeds from minimum speed up to 100% at intervals of 10%. This function is the same as listed in "Load and Run helloRobot.ino" (page 88)

Example 7-12. Functions to rotate the robot

```
void moveRotate(int angle)
{
  Serial.print("Rotating ");  Serial.println(angle);
  if(angle < 0)
  {
    Serial.println(" (left)");
    motorReverse(MOTOR_LEFT,  moveSpeed);
    motorForward(MOTOR_RIGHT, moveSpeed);
    angle = -angle; changeMoveState(MOV_ROTATE);
  }
  else if(angle > 0)
  {
    Serial.println(" (right)");
    motorForward(MOTOR_LEFT,  moveSpeed);
    motorReverse(MOTOR_RIGHT, moveSpeed);
    changeMoveState(MOV_ROTATE);
  }
  int ms = rotationAngleToTime(angle, moveSpeed);
  movingDelay(ms);
  moveBrake();
}

// return the time in milliseconds to turn the given angle at the given speed
long rotationAngleToTime( int angle, int speed)
{
int fullRotationTime; // time to rotate 360 degrees at given speed

  if(speed < MIN_SPEED)
    return 0; // ignore speeds slower then the first table entry

  angle = abs(angle);

  if(speed >= 100)
    fullRotationTime = rotationTime[NBR_SPEEDS-1]; // the last entry is 100%
  else
  {
    int index = (speed - MIN_SPEED) / SPEED_TABLE_INTERVAL; // index into speed and time tables
    int t0 =  rotationTime[index];
    int t1 = rotationTime[index+1];    // time of the next higher speed
    fullRotationTime = map(speed,  speedTable[index],  speedTable[index+1], t0, t1);
```

```
  // Serial.print("index= ");   Serial.print(index); Serial.print(", t0 = "); Serial.print(t0);
  // Serial.print(", t1 = ");   Serial.print(t1);
 }
 // Serial.print(" full rotation time = ");   Serial.println(fullRotationTime);
  long result = map(angle, 0,360, 0, fullRotationTime);
  return result;
}




// rotate the robot from MIN_SPEED to 100% increasing by SPEED_TABLE_INTERVAL
void calibrateRotationRate(int direction, int angle)
{
  Serial.print(locationString[direction]);
  Serial.println(" calibration" );
  for(int speed = MIN_SPEED; speed <= 100; speed += SPEED_TABLE_INTERVAL)
  {
    delay(1000);
    //blinkNumber(speed/10);

    if( direction == DIR_LEFT)
    {    // rotate left
      motorReverse(MOTOR_LEFT,  speed);
      motorForward(MOTOR_RIGHT, speed);
    }
    else if( direction == DIR_RIGHT)
    {    // rotate right
      motorForward(MOTOR_LEFT,  speed);
      motorReverse(MOTOR_RIGHT, speed);
    }
    else
        Serial.println("Invalid direction");


    int time = rotationAngleToTime(angle, speed);

    Serial.print(locationString[direction]);
    Serial.print(": rotate ");              Serial.print(angle);
    Serial.print(" degrees at speed ");   Serial.print(speed);
    Serial.print(" for ");                  Serial.print(time);
    Serial.println("ms");
    delay(time);
    motorStop(MOTOR_LEFT);
    motorStop(MOTOR_RIGHT);
    delay(2000); // two second delay between speeds
  }
}
```

The moveRotate function will rotate a robot by the given angle. Negative angles turn counter clockwise, positive angles turn clockwise. Rotation is achieved by running the motors in opposite directions (see "How Robots Move" (page 5)).

Higher-Level Movement Functions

Higher level movement functions work together to provide a simple way to instruct the robot briefly move away from one obstacle while checking to see if it needs to avoid another obstacle encountered while taking evasive action.

Timed Move

Moves the robot in a specified direction for a specified duration.

Moving Delay

Checks for obstacles while delaying for a specified period. Uses checkMove ment() function in the Look module to see if an obstacle is detected in the current direction of movement.

Example 7-13. Higher level movement functions

```
/************* high level movement functions ****************/

//moves in the given direction at the current speed for the given duration in milliseconds
void timedMove(int direction, int duration)
{
  Serial.print("Timed move ");
  if(direction == MOV_FORWARD) {
    Serial.println("forward");
    moveForward();
  }
  else if(direction == MOV_BACK) {
    Serial.println("back");
    moveBackward();
  }
  else
    Serial.println("?");

  movingDelay(duration);
  moveStop();
}

// check for obstacles while delaying the given duration in ms
void movingDelay(long duration)
{
  long startTime = millis();
  while(millis() - startTime < duration) {
    // function in Look module checks for obstacle in direction of movement
    if(checkMovement() == false) {
      if( moveState != MOV_ROTATE) // rotate is only valid movement
      {
          Serial.println("Stopping in moving Delay()");
          moveBrake();
```

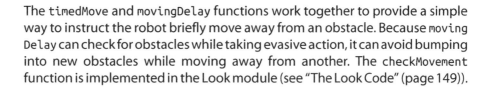

```
            }
          }
        }
      }
```

The timedMove and movingDelay functions work together to provide a simple way to instruct the robot briefly move away from an obstacle. Because moving Delay can check for obstacles while taking evasive action, it can avoid bumping into new obstacles while moving away from another. The checkMovement function is implemented in the Look module (see "The Look Code" (page 149)).

Tutorial: Introduction to Sensors

8

Sensor information can be used by your robot to navigate and interact with its environment. Sensors report on the world around them; measuring light, distance, sound, movement, direction, temperature, pressure, or location. This chapter describes how common sensors used with two wheeled and four wheeled platforms work.

The first half of this chapter covers the primary sensors used in the chapters that follow: IR reflective sensors and SONAR distance sensors. These are used to determine if an object is near the robot. Reflective sensors detect nearby objects and are used for line following and edge detection (determining if the robot is near the edge of the surface it is moving on, such as the edge of a table). Distance sensors are used to determine the distance to objects up to ten feet away from the robot. The second half of the chapter covers other types of sensors you can add to enable the robot to respond to distance, sound, movement, or other stimuli. You should also have a look at Appendix D, *Power Sources* which describes a very useful aspect to sense, the robot's battery voltage.

Hardware Discussed

QTR-1A reflectance sensors

Two are used for edge detection, but a third is required for line following. Additional sensors are available from many internet shops that stock robot parts, or direct from the manufacturer: *http://www.pololu.com/catalog/product/958/*

SONAR Distance Sensor

One is used to measure the distance to obstacles (Maker Shed product code MKPX5).

Maxbotix EZ1 distance sensor
> This is an optional item that can be used to measure distance.

Sharp IR
> This is an optional item that can be used to measure distance.

PIR (Passive Infrared) sensor
> This is an optional item that can be used to activate the robot when it detects the presence of a 'warm body' (Maker Shed product code: MKPX6).

Sound Sensor
> This is an optional item that can activate the robot on a sound level, such as a hand clap. (SparkFun product code BOB-09964).

Software

The chapter contains background information on sensors that will be added to the robot in later chapters. The reflectance sensor code is from the sketches introduced in Chapter 6, *Testing the Robot's Basic Functions*. The Ping (Sonar distance sensor) hardware and software is covered in Chapter 10, *Autonomous Movement*.

Infrared Reflectance Sensors

These sensors use reflected infrared light to detect the presence of a line for line following, or the absence of a reflection for edge (cliff) detection.

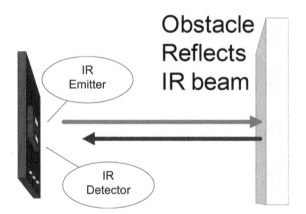

Figure 8-1. *Sensor using Infrared to detect obstacles*

The robot uses a function named `irSensorDetect`, shown in Example 8-1 to return true if the light level has increased sufficiently above the ambient level indicating that a nearby object is reflecting the IR beam.

Example 8-1. **Detecting an obstacle that reflects light**

```
const byte NBR_SENSORS = 3;  // this version only has left and right sensors
const byte IR_SENSOR[NBR_SENSORS] = {0, 1, 2}; // analog pins for sensors

// returns true if an object is detected on the given sensor
// the sensor parameter is the index into the sensor array
int irSensorDetect(int sensor)
{
  boolean result = false; // default value
  int value = analogRead(IR_SENSOR[sensor]); // get IR light level
  if( value <= irSensorReflect[sensor]) {
    result = true; // object detected (lower value means more reflection)
    if( isDetected[sensor] == false) { // only print on initial detection
      Serial.print(locationString[sensor]);
      Serial.println(" object detected");
    }
  }
  isDetected[sensor] = result;
  return result;
}
```

Sensor constants determine which sensor to use: SENSE_IR_LEFT for the left sensor, SENSE_IR_RIGHT for the right (these constants are defined in robotDe fines.h (see "Making the Sketch Easy to Enhance" (page 99)). The irSensor Detect function uses the sensor constant to retrieve the analog pin number stored in the IR_SENSOR array. If the analogRead value is less than a predeter-mined threshold, the function returns true indicating that a reflection has been detected. These functions use arrays instead of simple variables to store pins and thresholds because arrays make it easy to extend the code to support any number of sensors. To add a sensor, increase the NBR_SENSORS constant and add the sensors pin number to the list of pins in the IR_SENSOR array.

> *The sensor voltage reduces with increased light, so lower readings mean more reflectance. Therefore, the closer a reflecting object is to the sensor, the lower the reading on the analog pin monitoring the sensor.*

Whereas irSensorDetect returns true when a reflection is detected, sometime you want the opposite case—to return true if an edge (no reflection) is detec-ted, as in Example 8-2. The irEdgeDetect provides this capability; it is used to return true when an edge is detected. In other words, when the sensor is looking downwards, no reflection from the surface is detected because a dark object is blocking the reflection or the nearest surface—probably the floor—is many inches away! This effect is used in the examples from Chapter 6 to detect when you've placed a dark object under the sensor.

Example 8-2. **Detecting the absence of a reflection**

```
boolean irEdgeDetect(int sensor)
{
  boolean result = false; // default value
  int value = analogRead(IR_SENSOR[sensor]); // get IR light level
  if( value >= irSensorEdge[sensor]) {
    result = true; // edge detected (higher value means less reflection)
    if( isDetected[sensor] == false) { // only print on initial detection
      Serial.print(locationString[sensor]);
      Serial.println(" edge detected");
    }
  }
  isDetected[sensor] = result;
  return result;
}
```

The sensors need to be calibrated to take ambient light into account. Reflectance sensors respond to sunlight and artificial light so a threshold is measured with no object near the sensor. Levels above this threshold mean the light level is *above ambient*, which indicates that a nearby object is reflecting the IR light from the sensor. Ambient light calibration is done using the code shown in Example 8-3.

Example 8-3. **Light calibration**

```
// calibrate thresholds for ambient light
void irSensorCalibrate(byte sensor)
{
   int ambient = analogRead(IR_SENSOR[sensor]); // get ambient level
  irSensorAmbient[sensor] = ambient;
  // precalculate the levels for object and edge detection
  irSensorReflect[sensor] = (ambient * (long)(100-irReflectThreshold)) / 100;
  irSensorEdge[sensor]    = (ambient * (long)(100+irEdgeThreshold)) / 100;
}
```

> *(long) is used in the calculation to prevent overflow. Values like 95000*
> *cannot fit into an Arduino integer (max value is 32,767) whereas a long*
> *can store values up to 2,147,483,647.*
>
> *You may come across code that performs this calculation using floating*
> *point (ambient * 0.95). However, floating point requires more code*
> *and memory than integer calculations.*

This loads the ambient light level into the variable ambient, calculates levels for reflectance detection (stored in the irSensorReflect array) and levels for edge detection (stored in the iresensorEdge array). The constant irReflect Threshold is the percentage difference in light to detect a reflecting obstacle. The constant iredgeThreshold is the percent difference to detect an edge. The default values for these thresholds are 10% for reflection and 90% for edge detection.

Here is an example assuming the ambient value from analogRead was 1000 with irReflectThreshold equal to 10 :

```
(1000 * 90) / 100 =
90000 / 100 =
900
```

In this example, if the ambient reading was 1000, the irSensorReflect's threshold reading for object detection is 900, which is 10% below the ambient reading.

Sonar Distance Sensors

Sound pulses can be used to measure distance. The time it takes for a pulse to bounce off an object and return to the sensor is proportional to the distance.

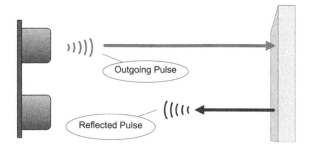

Figure 8-2. *Ping Sensor using SONAR to determine distance*

The speed of sound is 340 meters per second, which means it takes 29 micro-seconds for sound to travel 1 centimeter (the reciprocal of 340 metres per second). To derive the distance in cm, the duration is divided by 29. The duration is the time for the sum of outgoing and reflected pulses so the distance to the object is `microseconds / 29 / 2`.

The pulse duration is measured using the Arduino `pulseIn` function. This returns the a pulse duration in microseconds, see *http://arduino.cc/en/Reference/pulseIn*.

Example 8-4 shows the code that uses the Ping sensor to return the distance in inches. You'll see this code in action in Chapter 10.

Example 8-4. The source code for the Ping sensor

```
/*********************************
  code for ping distance sensor
*********************************/

// Returns the distance in inches
// this returns 0 if no ping sensor is connected or the distance is greater than around 10 feet
int pingGetDistance(int pingPin)
{
  // establish variables for duration of the ping,
  // and the distance result in inches and centimeters:
  long duration, cm;

  // The PING))) is triggered by a HIGH pulse of 2 or more microseconds.
  // Give a short LOW pulse beforehand to ensure a clean HIGH pulse:
  pinMode(pingPin, OUTPUT);
  digitalWrite(pingPin, LOW);
  delayMicroseconds(2);
  digitalWrite(pingPin, HIGH);
  delayMicroseconds(5);
  digitalWrite(pingPin, LOW);

  pinMode(pingPin, INPUT);
  duration = pulseIn(pingPin, HIGH, 20000); // if a pulse does not arrive
                                            // in 20 ms then the ping sensor
                                            // is not connected
  if(duration >=20000)
    return 0;

  // convert the time into a distance
  cm = microsecondsToCentimeters(duration);
  return (cm * 10) / 25 ; // convert cm to inches
}

long microsecondsToCentimeters(long microseconds)
{
  // The speed of sound is 340 m/s or 29 microseconds per centimeter.
```

```
// The ping travels out and back, so to find the distance of the
// object we take half of the distance travelled.
return microseconds / 29 / 2;
}
```

The pingGetDistance function returns the distance in inches as measured with a ping sensor on the digital pin (pingPin) passed to the function. The sound pulse used to measure the distance is triggered by sending a digital pulse that is low for 2 microseconds and high for 5 microseconds. The pin mode is changed from output to input and the pulseIn function is used to measure the response from the sensor, which arrives as an incoming pulse width. The formula described at the beginning of this section is used to convert this value to the distance.

Maxbotix EZ1 Sonar Distance Sensor

Example 8-5 shows the code for the Maxbotix EZ1 SONAR distance sensor (pictured in Figure 8-3).

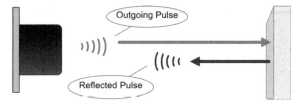

Figure 8-3. *Maxbotix EZ1 SONAR distance sensor*

Example 8-5. *Code for the EZ1 sensor*
```
/********************************
  code for the EZ1 SONAR sensor
*********************************/
// return distance using EZ1 connected to analog pin
int ezDistanceAN(int pin) // using analog
{
   const int bitsPerInch = 2; // each bit is 0.5 inch
   int value = analogRead(pin);
   int inches = value / 2;
   return inches;
}

// return distance using EZ1 connected using PW
int ezDistancePW(int pin) //using digital pin
{
   int value = pulseIn(pin, HIGH); // timeout can be added (MAX_DISTANCE * 147L * 2)
```

```
int cm = value / 58;      // pulse width is 58 ms per cm
int inches = value / 147; // which is 147 ms per inch
return inches;
}
```

The version using pulse width (ezDistancePW) will wait for one second before giving up if no return pulse is detected (for example, if the sensor is disconnected). You can optionally set the maximum time to wait for pulseIn; the following example sets the timeout to the duration needed for a pulse to travel the maximum distance detectable by the sensor:

```
int value = pulseIn(pin, HIGH, MAX_DISTANCE * 147L * 2);
// pulseIn with timeout
```

You can use the analog input version (ezDistanceAN), if you have a spare analog input pin, but if you only have digital pins free, then use the pulse width code (ezDistancePW). The analog version takes only as long as needed to measure the voltage so does not need a timeout.

You can find more information on this sensor on the manufacturers web page: *http://www.maxbotix.com/Ultrasonic_Sensors/MB1010.htm* .

Sharp IR Distance Sensor

Example 8-6 shows the code for the Sharp GP2Y0A02YK0F long range IR distance sensor (pictured in Figure 8-4).

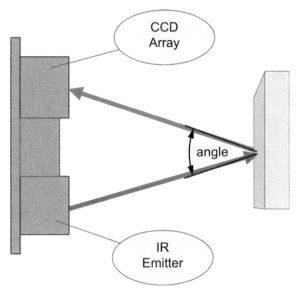

Figure 8-4. *Sharp IR Distance Sensor*

Example 8-6. Code for the Sharp IR sensor

```
/*********************************
 code for Sharp GP2Y0A02YK0F IR distance sensor
*********************************/
const long referenceMv = 5000; // the reference voltage in millivolts

int irGetDistance(byte pin)
{
  int val = analogRead(pin);
  int mV = map(val, 0, 1023, 0 , referenceMv);
  // or:
  //int mV = (val,  * referenceMv) / 1023;

  int cm = mvToDistance(mV);
  return cm;
}

// the following is used to interpolate the distance from a table
// table entries are distances in steps of 250 millivolts
const int TABLE_ENTRIES = 11;
const int firstElement = 250; // first entry is 250 mV
```

```
const int INTERVAL  = 250; // millivolts between each element
static int distance[TABLE_ENTRIES] = {200,130,90,64,50,41,35,30,25,20,15};

int mvToDistance(int mV)
{
    if( mV < firstElement )
        return distance[0];
    if( mV >  INTERVAL * TABLE_ENTRIES )
        return distance[TABLE_ENTRIES-1];
    else
    {
        int index = mV / INTERVAL;    // highest table element <= mV value
        int mV0 =  index * INTERVAL; // mV value of this element
        int mV1 = mV0 + INTERVAL;     // mV value of the next higher element
        int result = map(mV, mV0, mV1, distance[index-1], distance[index]);
        result = map(result, 0, 200, 0, 79); // convert from cm to inches
        return result;
    }
}
```

You can find lots more information on this sensor here: *http://www.societyo frobots.com/sensors_sharpirrange.shtml* .

Proximity Sensor

A PIR (Passive Infrared) sensor can be used to activate your robot when it detects the presence of a nearby person, or even a dog or cat. The sensor acts like a switch that sends a HIGH signal to an Arduino pin when motion is detected (they work by detecting changes in the heat radiated from people or pets). Figure 8-5 shows the sensor connected to analog pin 5, but you can use any spare pin, such as A4 instead of A5.

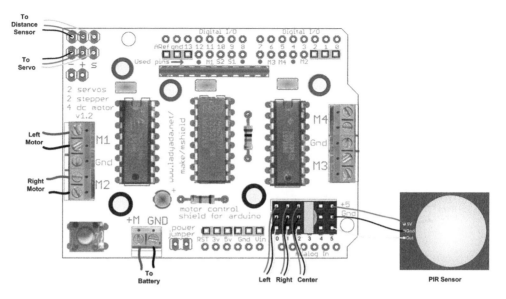

Figure 8-5. *PIR Sensor Connected to Analog Pin 5*

The following `loop` code will spin the robot when movement is detected. If you want your robot to do this, replace the `loop` function in the `myRobot` sketch from "Making the Sketch Easy to Enhance" (page 99) with the code shown in Example 8-7.

Example 8-7. Spinning the bot

```
void loop()
{
  Serial.println("Waiting to detect movement from PIR sensor");
  pinMode(A5, INPUT); // configure the pin for input
  if(digitalRead(A5) == HIGH)
  {
    calibrateRotationRate(DIR_LEFT, 360);  // spin robot CCW one rotation
  }
}
```

Sound Sensor

You can use a sound sensor to start or stop your robot in response to sound, for example a hand clap or whistle. You will need a microphone with an amplifier, for example, the BOB-09964 breakout board from SparkFun. Figure 8-6 shows the board connected to analog pin 4.

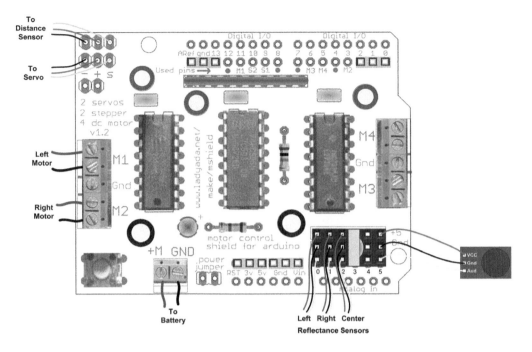

Figure 8-6. *Sound Sensor Connected to Analog Pin 4*

The code that follows is the main tab from the myRobotSound sketch available in the download for this book. Noise level above a threshold will drive the robot forward. The robot stops when the level drops below the threshold. If you need to change the sensitivity, experiment with higher or lower values for the threshold. Example 8-8 shows the code for the main tab.

*Example 8-8. **Sound sensor code***

```
/**********************************************************
MyRobotSound.ino

Robot moves when a sound level exceeds a threshold
Based on Recipe 6.7 from Arduino Cookbook

Copyright Michael Margolis 20 July 2012

**********************************************************/

#include <AFMotor.h>  // adafruit motor shield library
#include "RobotMotor.h"    // 2wd or 4wd motor library

#include "robotDefines.h"  // global defines

const int analogInPin = 5;      // analog pin the sensor is connected to
```

```
const int middleValue = 512;      //the middle of the range of analog values
const int numberOfSamples = 128;  //how many readings will be taken each time

int sample;                       //the value read from microphone each time
long signal;                      //the reading once you have removed DC offset
long averageReading;              //the average of that loop of readings

long runningAverage=0;            //the running average of calculated values
const int averagedOver= 16;       //how quickly new values affect running average
                                  //bigger numbers mean slower

const int threshold=400;          //at what level the robot will move

int speed = 50;

// Setup runs at startup and is used configure pins and init system variables
void setup()
{
  Serial.begin(9600);
  blinkNumber(8); // open port while flashing. Needed for Leonardo only

  motorBegin(MOTOR_LEFT);
  motorBegin(MOTOR_RIGHT);
}

void loop()
{
  int level = getSoundLevel();
  if (level > threshold)          //is level more than the threshold ?
  {
     motorForward(MOTOR_LEFT, speed);
     motorForward(MOTOR_RIGHT, speed);
  }else
  {
     motorStop(MOTOR_LEFT);
     motorStop(MOTOR_RIGHT);
  }
}

// function to indicate numbers by flashing the built-in LED
void blinkNumber( byte number) {
   pinMode(LED_PIN, OUTPUT); // enable the LED pin for output
   while(number--) {
     digitalWrite(LED_PIN, HIGH); delay(100);
     digitalWrite(LED_PIN, LOW);  delay(400);
   }
}

int getSoundLevel()
{
  long sumOfSquares = 0;
  for (int i=0; i<numberOfSamples; i++) { //take many readings and average them
    sample = analogRead(analogInPin);    //take a reading
    signal = (sample - middleValue);     //work out its offset from the center
```

```
    signal *= signal;                    //square it to make all values positive
    sumOfSquares += signal;              //add to the total
  }
  averageReading = sumOfSquares/numberOfSamples;     //calculate running average
  runningAverage=(((averagedOver-1)*runningAverage)+averageReading)/averagedOver;

  return runningAverage;
}
```

See the Arduino Cookbook if you want a detailed description of how this code works.

Arduino Cookbook

For descriptions of how to use lots of additional sensors with Arduino, see: *Arduino Cookbook* by Michael Margolis (O'Reilly).

Modifying the Robot to React to Edges and Lines

8

This chapter covers techniques that enable your robot to use sensors to gain awareness of its environment. Using reflectance sensors, the robot will gain the ability to follow lines or to avoid falling off the edge of the surface it is on. Information from the sensors is abstracted so that the robot logic has a single consistent interface and can easily be enhanced to support other sensors. The physical mounting of the sensors varies with different platforms: see Chapter 4, *Building the Four-Wheeled Mobile Platform* if you have the 4WD chassis, Chapter 3, *Building the Two-Wheeled Mobile Platform* if you have the 2WD chassis.

Hardware Required

- Two reflectance sensors are used for edge detection and a third is needed for line following. Although you can use the stripboard mount (for the three line following sensors) discussed in Chapter 2 to experiment with edge detection, the robot will perform the edge detection task best with the sensors further apart (the stripboard approach is best for line following). If the sensors are close together, the robot can have difficulty determining the best angle to turn when an edge is encountered.

> See *Chapter 3, Building the Two-Wheeled Mobile Platform* for details of mounting these sensors on the 2WD chassis and *Chapter 4, Building the Four-Wheeled Mobile Platform* for the 4WD chassis. The principles of reflectance sensors are covered in *"Infrared Reflectance Sensors" (page 134)* in Chapter 8, Tutorial: Introduction to Sensors.

- A reflective surface with non-reflective edges for the edge detection sketch (see Figure 9-2. You can use a large sheet of plain white paper with the edges marked using a black marker pen or black electrical tape. The border should be around 3/4 of in inch thick or so. The optimal surface would be white, but with sufficient friction that your robot won't slip. Lining paper, often sold as unpasted wall liner, is a great surface. It's designed to provide an even surface or wallpapering or painting, but with enough texture that makes it great for racing robots.

- A reflective surface with a non-reflective line approximately 3/4 inch wide, see Figure 9-3. If your surface is at least a couple of feet wide, you can use the same course for edge detection and line following. The sketches have been tested using a three foot length of 27 inch wide lining paper.

Sketches Used in This Chapter

myRobotEdge.ino

The robot will move about in an area bounded by a non-reflective surface (for example, a large sheet of white paper placed on a non-reflective surface. Most surfaces that reflect visible light will reflect infrared from the sensor).

myRobotLine.ino

This repositions the sensors used in myRobotEdge.ino to allow the robot to follow black lines painted on or taped to a white surface. The only change to the tab code is support for the center sensor. A variant of this sketch that sends data over serial for display on an external serial device is named myRobotLineDisplay and is included in the example code download (see "How to Contact Us" (page xv)).

Figure 9-1 shows the organization of the modules for this chapter.

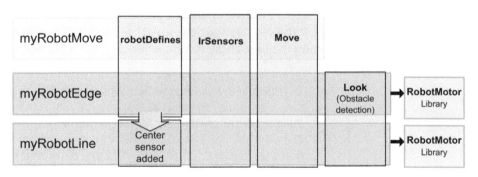

Figure 9-1. *myRobotEdge and myRobotLine Sketches*

The Look Code

The code to look for an obstacle and return `true` if detected is implemented in the function named `lookForObstacle`. You saw this function in the main tab of the sketch described in Chapter 6, *Testing the Robot's Basic Functions*. Because this code will be extended in this and later chapters to support additional sensors, it makes sense to extract this code into its own tab. The download code for all sketches introduced from here on in have a tab named `Look` that contains the code shown in Example 9-1.

Example 9-1. **Code for the Look tab**

```
/***********************
 code to look for obstacles
***********************/

void lookBegin()
{
  irSensorBegin();    // initialize sensors
}

// returns true if the given obstacle is detected
boolean lookForObstacle(int obstacle)
{
  switch(obstacle) {
     case  OBST_FRONT_EDGE: return irEdgeDetect(SENSE_IR_LEFT) && irEdgeDetect(SENSE_IR_RIGHT);
     case  OBST_LEFT_EDGE:  return irEdgeDetect(SENSE_IR_LEFT);
     case  OBST_RIGHT_EDGE: return irEdgeDetect(SENSE_IR_RIGHT);
  }
  return false;
}

// function to check if robot can continue moving when taking evasive action
// returns true if robot is not blocked when moving to avoid obstacles
// this 'placeholder' version always returns true
boolean checkMovement()
{
  return true;
}
```

As mentioned in Chapter 6, *Testing the Robot's Basic Functions*, the `lookForObstacle` function enables you to enquire if an obstacle is detected and will return true if so. The case statement (see *http://arduino.cc/en/Reference/SwitchCase*) tries to match the `obstacle` variable with one of the obstacle constants (defined in `robotDefines.h`). If there is a match, the `irEdgeDetect` function is called with relevant sensor and this will return true if an object is detected on that sensor. If no object is detected, the function returns OBST_NONE. The look functionality can be expanded by adding code to the case statement and calling appropriate sensor functions, as you will see later in this chapter.

But first, let's use the existing functionality to give the robot the ability to follow lines and detect edges.

Edge Detection

Edge detection is one of the easier behaviors to understand and program. The robot moves until it encounters an edge; it should then change direction to avoid moving over the edge. Edges are detected by using reflectance sensors (see: Chapter 8, *Tutorial: Introduction to Sensors*). Typically, the edge is an area that does not reflect, for example the edge of a table.

In the sketch that follows, the robot will remain within a reflective surface (for example, a large white sheet of paper) that is bounded by a black line. Black electrical tape (3/4 inch or wider) works well but a black line of similar width drawn with magic marker or paint can also work as the 'edge'. To avoid damaging your robot, an actual table is not recommended for early experiments until you are sure you have everything working correctly.

Figure 9-2 shows how the robot responds to moving over an edge. In panel1, the sensors do not detect an edge so the robot moves forward. In panel 2, the left sensor moves off the reflective surface so the robot stops and rotates 120 degrees. In panel 3, the robot completes its rotation; panel 4, shows the robot moving forward again.

Figure 9-2. *Robot stays within the reflective area*

Example 9-2. **Main sketch code for edge detection**

```
/************************************************************************
myRobotEdge.ino

Robot sketch to move within area bordered by a non-reflective line

Michael Margolis 7 July 2012
************************************************************************/
```

```
#include <AFMotor.h>  // adafruit motor shield library
#include "RobotMotor.h"    // 2wd or 4wd motor library

#include "robotDefines.h"  // these were the global defines from myRobot

/// Setup runs at startup and is used configure pins and init system variables
void setup()
{
  Serial.begin(9600);
  blinkNumber(8); // open port while flashing. Needed for Leonardo only

  lookBegin();  /// added Look tab
  moveBegin();  /// added Move tab
  Serial.println("Ready");
}

void loop()
{
  /// code for roaming around and avoiding obstacles
  if( lookForObstacle(OBST_FRONT_EDGE) == true)
  {
    Serial.println("both sensors detected edge");
    timedMove(MOV_BACK, 300);
    moveRotate(120);
    while(lookForObstacle(OBST_FRONT_EDGE) == true )
      moveStop(); // stop motors if still over cliff
  }
  else if(lookForObstacle(OBST_LEFT_EDGE) == true)
  {
    Serial.println("left sensor detected edge");
    timedMove(MOV_BACK, 100);
    moveRotate(30);
  }
  else if(lookForObstacle(OBST_RIGHT_EDGE) == true)
  {
    Serial.println("right sensor detected edge");
    timedMove(MOV_BACK, 100);
     moveRotate(-30);
  }
  else
  {
    moveSetSpeed(MIN_SPEED);
    moveForward();
  }
}

// function to indicate numbers by flashing the built-in LED
void blinkNumber( byte number) {
   pinMode(LED_PIN, OUTPUT); // enable the LED pin for output
   while(number--) {
     digitalWrite(LED_PIN, HIGH); delay(100);
```

```
    digitalWrite(LED_PIN, LOW);  delay(400);
  }
}
```

The code for this sketch is derived from the myRobotMove sketch discussed in Chapter 7, *Controlling Speed and Direction*. You can download the example code, locate myRobotEdge, open the sketch and upload it to the robot. Or you can derive the sketch yourself:

1. Open the myRobotMove sketch in the example code and do a Save As and name it myRobotEdge.

2. Create the Look tab.

3. Locate and move the two functions at the end of the main tab starting from the comment "code to look for obstacles" into the Look tab. This code is listed in the section: "The Look Code" (page 149).

4. Replace the main sketch code with the code listed here: Example 9-2.

5. Compile and upload the code

Place the robot within the bounded surface and switch the power on (the robot calibrates the sensors after it is switched on so all the sensors should be over the reflective area). After a short delay the robot will move forward until it detects a non-reflective edge.

The loop code checks if an edge is detected directly ahead with both sensors (OBST_FRONT_EDGE), or on the left (OBST_LEFT_EDGE) or right (OBST_RIGHT_EDGE). If the edge was ahead, the robot backs away for 0.3 seconds, rotates 120 degrees and then moves forward again. If the edge was to the side, the robot turns 30 degrees away from that side and then moves forward. Feel free to experiment with the angles to get a behaviour that suits the area you have defined for containing your robot.

Is Your Robot Not Moving Right?

If your robot is not rotating enough or too much when attempting to move away from an edge, you may need to calibrate rotation rates; see "Controlling Motor Speed" (page 109). If the robot 'stutters' instead of turning, try increasing the speed by changing the loop code from moveSetSpeed(MIN_SPEED); to moveSetSpeed(MIN_SPEED +10);.

If the robot does not detect the edge, you can make it more sensitive by reducing the value of irEdgeThreshold in the IrSensors tab.

Line Following

Line following is a classic task for a robot. The robot uses sensors to determine its position in relation to a line and follows this line by moving to keep its sensors centered above the line. Figure 9-3 shows a robot moving around a track marked with a black line on a white surface.

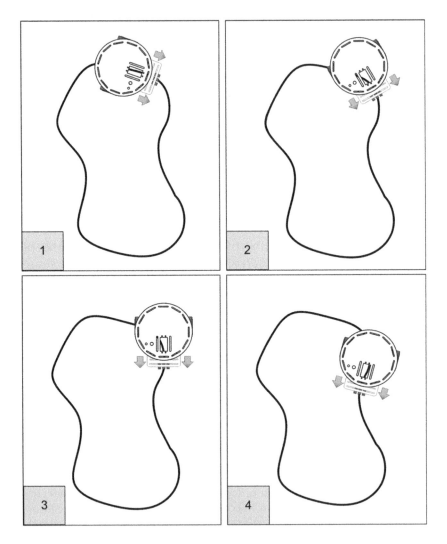

Figure 9-3. *Robot follows a black line on a white surface*

In Panel 1, the robot is approaching a corner but is still centered over the line - the motors are both running at the same speed (indicated by the equal length arrows), and the robot moves straight ahead. The robot has reached the left hand curve in panel 2—the right motor speed is increased, the left slowed to turn the robot to the right. Panel 3 shows the robot completing the turn. In Panel 4, the robot is about to reach a curve to the left where it will continue to adjust motor speeds to keep the sensors over the line.

The illustrations that follow show what happens in more detail. Figure 9-4 shows the location of the sensor with respect to the line when the robot is centered. The left and right sensors are above the reflective surface. Lots of light will reflect back to the sensor and the `analogRead` values are low. The center sensor is above the black line so has little reflected light, causing the reading to be high. The difference in readings between left and right indicates drift and is close to zero so both motors will be driven at the same speed—the robot moves straight ahead. You can read about how to display sketch data in real time in "Seeing Sketch Data" (page 160).

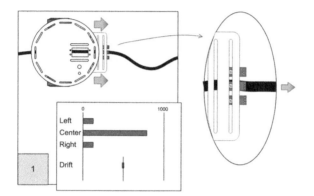

Figure 9-4. *Robot centered on black line*

Figure 9-5 shows the robot to the left of the line because the line is curving to the right. The left sensor detects maximum reflection (the `analogRead` value is low). As the center sensor moves towards the edge of the line, the reflection increases (decreasing the `analogRead` value). The right sensor moves towards the line so its reading increases. The drift (the difference between the left and right) is positive so the left motor speeds up and the right motor slows down —the robot turns to the right.

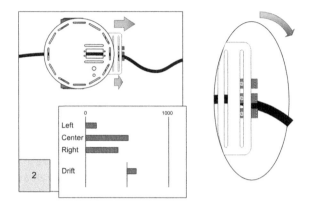

Figure 9-5. *Robot off to left of line*

Figure 9-6 shows the robot to the right of the line because the line is curving to the left. The right sensor detects maximum reflection (the `analogRead` value is low). The reading from the center sensor increases as it moves towards the edge of the line. The left sensor moves towards the line so its reading increases. The drift is negative so the left motor slows down and the right motor speeds up—the robot turns to the left.

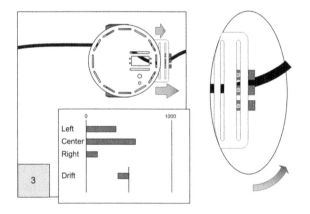

Figure 9-6. *Robot off to right of line*

For the robot to successfully follow a curvy line, the movement must be responsive enough to make sharp turns but not so responsive that it zigs and zags its way along even straight lines. Tuning the software to get this just right requires experimentation and patience. The code that follows uses the difference value between the left and right sensors to adjust the differential motor speed. The preceding figures display the relative signal levels from the sensors and the difference value is indicated as 'Drift'. Sensitivity is controlled by mapping the drift value to the actual motor differential speed.

Example 9-3 shows the line sense code that calculates the drift value (you'll see it again in a moment when you see the complete listing for the sketch):

Example 9-3. Line sense code for calculating drift
```
//returns drift - 0 if over line, minus value if left, plus if right
int lineSense()
{
  int leftVal = analogRead(SENSE_IR_LEFT);
  int centerVal = analogRead(SENSE_IR_CENTER);
  int rightVal = analogRead(SENSE_IR_RIGHT);

  int leftSense = centerVal - leftVal;
  int rightSense = rightVal - centerVal;
  int drift = rightVal - leftVal ;
  return drift;
}
```

The drift and desired speed are passed to the lineFollow function to drive the robot. To adjust the motor's sensitivity, drift is divided by a 'damping' factor - the higher the factor, the less sensitive to drift. Decrease the damping if you need to make the robot more sensitive, for example, if it is not turning fast enough to follow sharp bends. Increase the damping if the robot is unnecessarily zig-zagging on straight lines. The drift value is subtracted from the speed for the left motor and added to the speed of the right motor to provide a differential speed proportional to drift. The Arduino constrain function is used to ensure the values remain within the valid range for speed (0 to 100 %). Depending on the radius of your bends, you may not be able to completely eliminate the zig-zags.

```
int lineFollow(int drift, int speed)
{
  int leftSpeed  = constrain(speed - (drift / damping), 0, 100);
  int rightSpeed = constrain(speed + (drift / damping), 0, 100);

  motorForward(MOTOR_LEFT, leftSpeed);
  motorForward(MOTOR_RIGHT, rightSpeed);
}
```

Example 9-4. Complete listing for code in the myRobotLine main tab
```
/***************************************************************************
myRobotLine.ino

Robot sketch to follow lines

Michael Margolis 7 July 2012
***************************************************************************/
```

```
#include <AFMotor.h>  // adafruit motor shield library
#include "RobotMotor.h"    // 2wd or 4wd motor library

#include "robotDefines.h"  // these were the global defines from myRobot

int speed = MIN_SPEED; // speed in percent when moving along a straight line

/// Setup runs at startup and is used configure pins and init system variables
void setup()
{
  Serial.begin(9600);
  blinkNumber(8); // open port while flashing. Needed for Leonardo only

  lookBegin();  /// added Look tab
  moveBegin();  /// added Move tab
  lineSenseBegin();    // initialize sensors
  Serial.println("Ready");
}

void loop()
{
  int drift = lineSense();
  lineFollow(drift, speed);
}

// function to indicate numbers by flashing the built-in LED
void blinkNumber( byte number) {
    pinMode(LED_PIN, OUTPUT); // enable the LED pin for output
    while(number--) {
      digitalWrite(LED_PIN, HIGH); delay(100);
      digitalWrite(LED_PIN, LOW);  delay(400);
    }
}

/****************************
   Line   Sensor code
***************************/

int damping =  5; //1 is most sensitive, range 1 to 1023)

void lineSenseBegin()
{

}

//returns drift - 0 if over line, minus value if left, plus if right
int lineSense()
{
  int leftVal = analogRead(SENSE_IR_LEFT);
  int centerVal = analogRead(SENSE_IR_CENTER);
  int rightVal = analogRead(SENSE_IR_RIGHT);

  int leftSense = centerVal - leftVal;
  int rightSense = rightVal - centerVal;
```

```
   int drift = rightVal - leftVal ;
   return drift;
}

int lineFollow(int drift, int speed)
{
   int leftSpeed   = constrain(speed - (drift / damping), 0, 100);
   int rightSpeed  = constrain(speed + (drift / damping), 0, 100);

   motorForward(MOTOR_LEFT, leftSpeed);
   motorForward(MOTOR_RIGHT, rightSpeed);
}
```

The code for this sketch is derived from the myRobotEdge sketch discussed earlier in this chapter. You can download the example code, locate myRobot-Line, open the sketch and upload it to the robot. Or you can derive the sketch yourself:

1. Open the myRobotEdge sketch in the example code and do a Save As and name it myRobotLine.

2. Locate the defines for locations of sensors in the robotDefines tab and add the center sensor following the defines for the left and right sensors: const int SENSE_IR_CENTER = 2;.

3. Replace the main sketch code with the code listed here: Example 9-4.

4. Compile and upload the code

Place the robot on the surface with the center sensor above the line and switch the power on. After a short delay the robot will move forward and track the line. The robots ability to follow the line depends on many factors:

- Line thickness - the optimum thickness depends on the spacing of the sensors. 3/4 inch works well with the robot built as described but you can experiment with different line widths and different sensor spacing.

- Sensor height above surface - the sensors are less sensitive when further from the surface- try using spacers to move the sensors closer to the surface.

- Speed - too slow and the robot may not have enough torque, too fast and the robot will overshoot the line. Try running at a speed around 10% above minimum speed if the robot appears to be sluggish - in the top of the main tab, change the code to : int speed = MIN_SPEED+10;.

- Robot over sensitive - if the robot follows the line but zig-zags excessively, increase the damping value in the line sensor code in the main tab. Try larger values until you find a range that works. Note that you may need a different damping value if you change the speed.
- Robot not sensitive enough - if the robot drifts off the line, decrease the damping value in the line sensor code in the main tab. Try smaller values until you find a range that works. Note that you may need a different damping value if you change the speed.

Seeing Sketch Data

Viewing the values of variables in real time makes it much easier to tune or debug your code. You can view values printed to the serial port on the Serial Monitor, but that can be difficult to read if these values are changing quickly. Appendix C describes how to use a Processing sketch to display data as bar charts, similar to the that shown earlier in this chapter (see Figure 9-4).

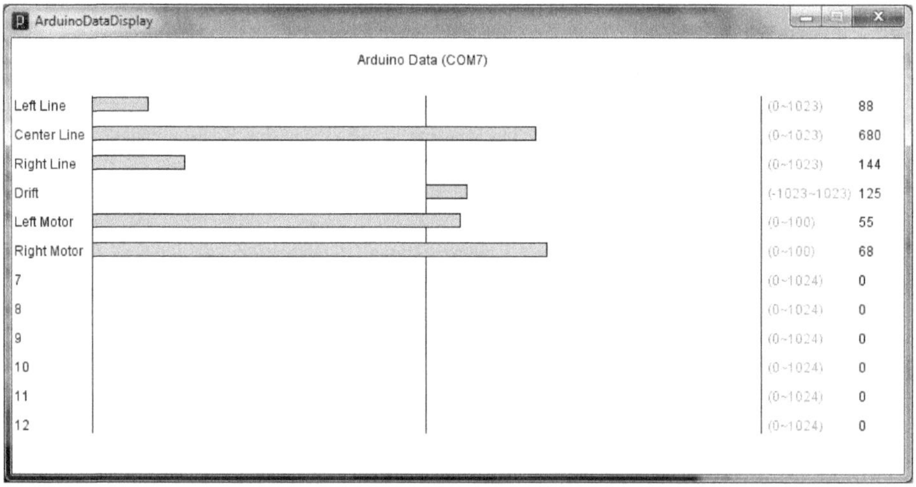

Figure 9-7. *Arduino Data Displayed in Processing*

Here is how the line following sketch should be modified to display the sensor value:

Add the DataDisplay tab to the sketch (the myRobotLineDisplay sketch in the example code download ("How to Contact Us" (page xv)) has this tab added as well as all the other code changes that follow).

In the main sketch, add constants identifying the list of items to be displayed, labels for each item, and the minimum and maximum values. Example 9-5 shows the constants used to produce the display in Figure 9-7.

Example 9-5. **Constants for display labels**

```
enum {DATA_start,DATA_LEFT,DATA_CENTER,DATA_RIGHT,
      DATA_DRIFT,DATA_L_SPEED,DATA_R_SPEED, DATA_nbrItems};

char* labels[] =
  {"", "Left Line", "Center Line", "Right Line","Drift", "Left Speed", "Right Speed"};

int minRange[] =
  {0,           0,              0,          0,  -1023,              0,              0};

int maxRange[] =
  {0,        1023,           1023,       1023,   1023,            100,            100};
```

Add the function call shown in Example 9-6 to setup().

Example 9-6. **Adding a call to begin the data display**

```
dataDisplayBegin(DATA_nbrItems, labels, minRange, maxRange );
```

You can then call the sendData function to send the values you want to display. Example 9-7 shows the lineSense() function updated to send sensor data.

Example 9-7. **lineSense now sending sensor data**

```
//returns drift - 0 if over line, minus value if left, plus if right
int lineSense()
{
  int leftVal = analogRead(SENSE_IR_LEFT);
  int centerVal = analogRead(SENSE_IR_CENTER);
  int rightVal = analogRead(SENSE_IR_RIGHT);

  sendData(DATA_LEFT, leftVal);       // send left sensor value
  sendData(DATA_CENTER, centerVal);   // send center sensor value
  sendData(DATA_RIGHT, rightVal);     // send right sensor values


  int leftSense = centerVal - leftVal;
  int rightSense = rightVal - centerVal;
  int drift = rightVal - leftVal ;
```

```
    sendData(DATA_DRIFT, drift);    // send drift sensor values

    return drift;
}
```

Motor speed can be displayed by adding calls to sendData in the lineFollow function as shown in Example 9-8.

*Example 9-8. **Adding support for displaying motor speed***

```
int lineFollow(int drift, int speed)
{
    int leftSpeed   = constrain(speed - (drift / damping), 0, 100);
    int rightSpeed  = constrain(speed + (drift / damping), 0, 100);

    sendData(DATA_L_SPEED, leftSpeed);    // send left motor speed
    sendData(DATA_R_SPEED, rightSpeed);   // send right motor speed

    motorForward(MOTOR_LEFT, leftSpeed);
    motorForward(MOTOR_RIGHT, rightSpeed);
}
```

Autonomous Movement 10

This chapter describes how to use a distance sensor to enable the robot to see and avoid obstacles as it moves around. The first sketch, named myRobotWander, drives the robot forward, and if it detects an obstacle, it stops and rotates the robot to try and find a clear path to move forward. Another sketch, named myRobotScan, adds a servo that can rotate the sensor so the robot can look left and right without having to twist itself around.

Hardware Required

- Ping distance sensor from Parallax; see "Sonar Distance Sensors" (page 137) in Chapter 8, *Tutorial: Introduction to Sensors*.

- Servo required for myRobotScan; see "Sonar Distance Sensors" (page 137) in Chapter 8, *Tutorial: Introduction to Sensors*.

Connect the Ping sensor and servo the right way around; the black wires (ground) go nearest the pin marked -, the white (or lighter color) signal wire goes nearest the pin marked **S** (Figure 10-1).

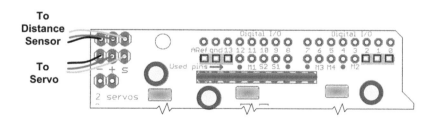

Figure 10-1. *Ping sensor and servo plug into pins on the motor shield*

Sketches Used in This Chapter

myRobotWander.ino

> Uses a SONAR distance sensor (the Ping sensor) to enable the robot to see and avoid obstacles as it wanders around. #defines are added for front and rear obstacles (only the front is implemented in the sketch), the look module has added support for distance sensing. This sketch introduces a new tab, named Distance, which contains the Ping sensor code that you originally saw in "Sonar Distance Sensors" (page 137).

myRobotScan.ino

> Has the sensor mounted on a servo so it can scan independently of robot movement. This code is similar to myrobotWander with the Look module enhanced to support control of the servo to look around. A new module named softServo is added for servo control.

> *The Distance tab's code is not listed in this chapter, but it is included in the example code (see **"How to Contact Us" (page xv)** for information on downloading the example code).*

Figure 10-2 shows the modules used in this chapter.

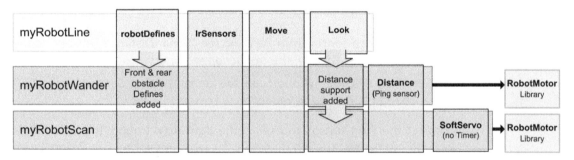

Figure 10-2. *myRobotWander and myRobotScan Sketches*

Mounting a Ping Distance Sensor

There are various ways to mount a Ping sensor. You can buy a commercial off-the-shelf product such as the one illustrated in Figure 10-3.

The bracket shown in Figure 10-4 and Figure 10-5 is another commercial off-the-shelf bracket that mounts the sensor on a servo so it can be rotated to scan for objects on either side of the robot. If you use a commercial off-the-shelf product, follow the supplied instructions for assembly and mounting.

Figure 10-3. *Parallax Ping Bracket*

Figure 10-4. *Parallax Ping Servo Bracket*

Figure 10-5. *Parallax servo bracket parts*

If you prefer to make a bracket, it's easy to do, as the next section explains.

Making a Mount for the Ping Sensor

You can make a simple mount from a small piece of wood. A small mount can be cut from 1" x 1 3/4" x 3/8" pine. You can drill holes for bolts (4-40 or M3) or use small wood screws to attach the sensor. The holes are close to the edges, so drill pilot holes if you use wood screws. Figure 10-6 shows a template you can use for the mount. You can see the Ping sensor attached to the mount in Figure 10-7 and Figure 10-8.

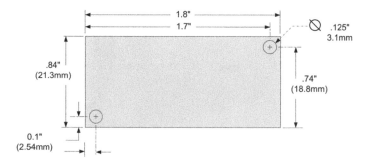

Figure 10-6. *Dimensions for the holes for a simple mount*

Figure 10-7 shows the a small block of wood cut and drilled.

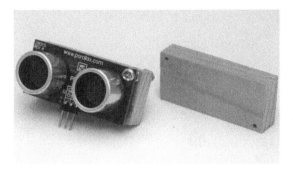

Figure 10-7. *Ping sensor with homemade wood mount, not fully assembled*

Figure 10-8. *Rear view of mount; note nuts used as spacers between PCB and mount*

Figure 10-9 and Figure 10-10 show the mount attached to the robot.

Figure 10-9. *Sensor can be mounted directly to the chassis or on a servo*

Figure 10-10. *Feel free to make your mount in a different size or shape*

Mounting the Ping Sensor in a Fixed Position

The manufactured mounts are supplied with mounting hardware. If you are using a homemade wooden mount, you can attach it to the base with two wood screws from the underside of the top plate.

Mounting the Ping Sensor on a Servo

The wooden mounts can be hot glued or screwed onto the servo horn supplied with the servo as shown in Figure 10-11. Figure 10-11 showed the sensor attached to a servo that's attached to the 2WD. Figure 10-12 shows the sensor on a servo attached to the 4WD.

Figure 10-11. *Servo mount showing attachment detail*

Figure 10-12. *2WD with sensor mounted onto servo*

Letting the Robot Wander

The myRobotWander sketch adds support for a fixed forward-facing distance sensor that enables the robot to move forward when no obstacle is detected (panel 1 in Figure 10-13). The robot stops when approaching an obstacle ahead (panel 2). It rotates left (panel 3) to see if there is an obstacle in that direction. If no obstacle is seen, the robot will turn in that direction and move off. If the left is obstructed, it will turn right and move off in that direction (panel 5). If both left and right are blocked, the robot will turn around and move off in the opposite direction.

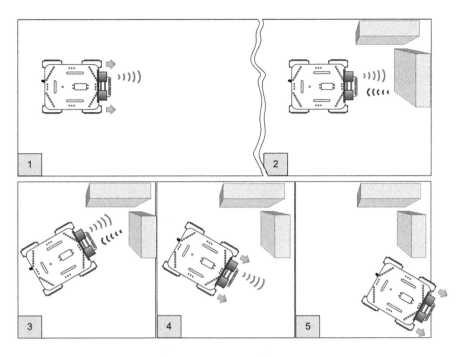

Figure 10-13. *Ping Sensor fixed in place to look for obstacles ahead*

The code to provide this behaviour is in the sketch named `myrobotWander`. Example 10-1 shows the main myRobotWander tab for this sketch.

Example 10-1. **Contents of the sketch's main tab**

```
/****************************************************************************
myRobotWander.ino

Robot wanders using forward scanning for obstacle avoidance

Michael Margolis 28 May 2012
****************************************************************************/

#include <AFMotor.h>  // adafruit motor shield library
#include "RobotMotor.h"    // 2wd or 4wd motor library

#include "robotDefines.h"  // global defines

// Setup runs at startup and is used configure pins and init system variables
void setup()
{
  Serial.begin(9600);
  blinkNumber(8); // open port while flashing. Needed for Leonardo only

  lookBegin();
  moveBegin();
```

```
    moveSetSpeed(MIN_SPEED + 10) ;  // Run at 10% above minimum speed
    Serial.println("Ready");
}

void loop()
{
    moveForward();
    roam();  // look around
}

// function to indicate numbers by flashing the built-in LED
void blinkNumber( byte number) {
    pinMode(LED_PIN, OUTPUT); // enable the LED pin for output
    while(number--) {
      digitalWrite(LED_PIN, HIGH); delay(100);
      digitalWrite(LED_PIN, LOW);  delay(400);
    }
}
```

This simply initializes the 'Look' module ("The Look Code" (page 149)) and 'Move' module ("Core Movement Code" (page 124)) and then calls a function named roam that does all the hard work of looking for obstacles and moving to avoid them. The roam function is added into code in the Look tab; Example 10-2 replaces the entirety of the Look tab code that you saw in earlier examples.

Example 10-2. *The new version of the Look tab code*
```
/***********************
  code to look for obstacles
 ***********************/

const int  MIN_DISTANCE = 8;      // robot stops when object is nearer (in inches)
const int  CLEAR_DISTANCE = 24;   // distance in inches considered attractive to move
const int  MAX_DISTANCE = 150;    // the maximum range of the distance sensor

// angles                left, right, center
const int lookAngles[] = {  -30,  30,   0};

const byte pingPin = 10; // digital pin 10

void lookBegin()
{
    irSensorBegin();    // initialize sensors
}

// returns true if the given obstacle is detected
boolean lookForObstacle(int obstacle)
{
```

```
  switch(obstacle) {
     case  OBST_FRONT_EDGE: return irEdgeDetect(DIR_LEFT) && irEdgeDetect(DIR_RIGHT);
     case  OBST_LEFT_EDGE:  return irEdgeDetect(DIR_LEFT);
     case  OBST_RIGHT_EDGE: return irEdgeDetect(DIR_RIGHT);
     case  OBST_FRONT:      return  lookAt(lookAngles[DIR_CENTER]) <= MIN_DISTANCE;
  }
  return false;
}

// returns the distance of objects at the given angle
// this version rotates the robot
int lookAt(int angle)
{
  moveRotate(angle);    // rotate the robot

  int distance, samples;
  long cume;
  distance = samples = cume = 0;
  for(int i =0; i < 4; i++)
  {
    distance = pingGetDistance(pingPin);
    if(distance > 0)
    {
      //  printlnValue(" D= ",distance);
      samples++;
      cume+= distance;
    }
  }
  if(samples > 0)
    distance = cume / samples;
  else
    distance = 0;

  moveRotate(-angle);    // rotate back to original direction
  return distance;
}

// function to check if robot can continue moving in current direction
// returns true if robot is not blocked moving in current direction
// this version only tests for obstacles in front
boolean checkMovement()
{
  boolean isClear = true; // default return value if no obstacles
  if(moveGetState() == MOV_FORWARD)
  {
    if(lookForObstacle(OBST_FRONT) == true)
    {
       isClear = false;
    }
  }
  return isClear;
}

// Look for and avoid obstacles by rotating robot
```

```
void roam()
{
  int distance = lookAt(lookAngles[DIR_CENTER]);
  if(distance == 0)
  {
    moveStop();
    Serial.println("No front sensor");
    return;  // no sensor
  }
  else if(distance <= MIN_DISTANCE)
  {
    moveStop();
    //Serial.print("Scanning:");
    int leftDistance  = lookAt(lookAngles[DIR_LEFT]);
    if(leftDistance > CLEAR_DISTANCE)  {
//    Serial.print(" moving left: ");
      moveRotate(-90);
    }
    else {
      delay(500);
      int rightDistance = lookAt(lookAngles[DIR_RIGHT]);
      if(rightDistance > CLEAR_DISTANCE) {
//    Serial.println(" moving right: ");
        moveRotate(90);
      }
      else {
       // Serial.print(" no clearence : ");
        distance = max( leftDistance, rightDistance);
        if(distance < CLEAR_DISTANCE/2) {
          timedMove(MOV_BACK, 1000); // back up for one second
          moveRotate(-180); // turn around
        }
        else {
          if(leftDistance > rightDistance)
            moveRotate(-90);
          else
            moveRotate(90);
        }
      }
    }
  }
}

// the following is based on loop code from myRobotEdge
// robot checks for edge and moves to avoid
void avoidEdge()
{
  if( lookForObstacle(OBST_FRONT_EDGE) == true)
  {
    Serial.println("left and right sensors detected edge");
    timedMove(MOV_BACK, 300);
    moveRotate(120);
    while(lookForObstacle(OBST_FRONT_EDGE) == true )
      moveStop(); // stop motors if still over cliff
```

```
    }
    else if(lookForObstacle(OBST_LEFT_EDGE) == true)
    {
        Serial.println("left sensor detected edge");
        timedMove(MOV_BACK, 100);
        moveRotate(30);
    }
    else if(lookForObstacle(OBST_RIGHT_EDGE) == true)
    {
        Serial.println("right sensor detected edge");
        timedMove(MOV_BACK, 100);
        moveRotate(-30);
    }
}
```

The roam function uses information reported by the distance sensor to detect obstacles. The distance sensor code is described in "Sonar Distance Sensors" (page 137), the sketches in this chapter contain the code in a new tab named Distance.

The checkMovement function introduced in the previous chapter is enhanced here to check for and return false if there are obstacles in front when the robot is moving forward. checkMovement is called when the robot is taking evasive action during a timed move. You can add additional checks into this function if needed. For example, if you add sensors to detect an edge to the rear of the robot and added your own code that returned true when this sensor detected an edge, the logic shown in Example 10-3 would prevent the robot from going over an edge when backing up to avoid an obstacle in front.

Example 10-3. **The checkMovement function**

```
boolean checkMovement()
{
    boolean isClear = true; // default return value if no obstacles
    if(moveGetState() == MOV_FORWARD)
    {
        if(lookForObstacle(OBST_FRONT) == true)
        {
            isClear = false;
        }
    }
    else if(moveGetState() == MOV_BACK)
    {
        if(lookForObstacle(OBST_REAR_EDGE) == true)
        {
            isClear = false;
```

```
      }
    }
    return isClear;
}
```

In this fragment, if the robot is moving backward a call is made to lookForOb stacle (with a new case you need to add for a rear edge sensor) that checks if an edge is detected at that back of the robot.

The rest of the Look code is similar to the code described in Chapter 9, *Modifying the Robot to React to Edges and Lines*. The lookForObstacle function has an additional case for detecting an obstacle in front (OBST_FRONT). This case calls a new function named lookAt that is given the angle to look towards, and returns the distance of the nearest object detected at that angle. That distance is compared to a minimum allowable distance and lookForObstacle returns true if the robot is any closer (in other words, it has detected an obstacle).

The lookAt function (repeated in Example 10-4 from the previous listing) rotates the robot to the desired angle using the moveRotate command described in Chapter 7, *Controlling Speed and Direction*.

Example 10-4. The lookAt function
```
// returns the distance of objects at the given angle
// this version rotates the robot
int lookAt(int angle)
{
  moveRotate(angle);    // rotate the robot

  int distance, samples;
  long cume;
  distance = samples = cume = 0;
  for(int i =0; i < 4; i++)
  {
    distance = pingGetDistance(pingPin);
    if(distance > 0)
    {
      samples++;
      cume+= distance;
    }
  }
  if(samples > 0)
    distance = cume / samples;
  else
    distance = 0;
```

```
  moveRotate(-angle);    // rotate back to original direction
  return distance;
}
```

The pingGetDistance function (Example 8-4) returns the distance in inches. To minimize spurious reflection affecting the readings, the function is called four times to get an average distance. After taking the readings, the robot is rotated so it is facing in the original direction. Because the robot doesn't rotate to exactly the angle requested (due to changes in battery voltage, friction, etc.), the robot may not end up facing exactly the same direction and may appear to zig-zag as it moves forward.

The #defines shown in Example 10-5 are added to the robotDefines tab.

Example 10-5. New constants for front and rear detection
```
const int OBST_FRONT    = 4;  // obstacle in front
const int OBST_REAR     = 5;  // obstacle behind
```

Adding Edge Detection

Unlike the earlier chapters, the sketches in this chapter do not try to avoid edges. This enables the robot to wander over surfaces such as wooden floors that could create false triggers on the edge sensors. The edge detection code that was in the loop of the myRobotEdge sketch has been moved to the Look tab into a function named avoidEdge. If you want to add edge detection capability to these sketches, add a call to avoidEdge in loop as follows:

```
void loop()
{
  moveForward();
  roam();       // look around
  avoidEdge(); // avoid edges
}
```

Adding Scanning

In the previous sketch, the robot needs to turn in order to look left and right. Mounting the distance sensor on a servo adds the ability to rotate the sensor so the robot can 'turn its head' to look around as shown in Figure 10-14.

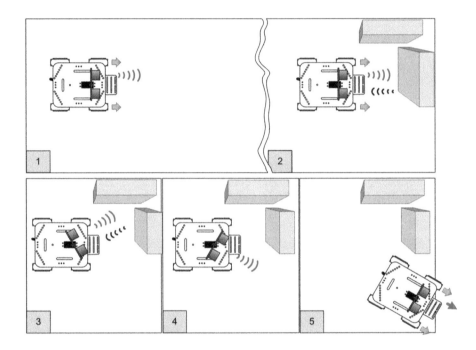

Figure 10-14. *Robot Scans using Ping Sensor Mounted on Servo*

The sketch logic is the same, but the Look module has code added to command a servo to rotate left and right for brief periods. This allows the sensor to look to see if it can detect an obstacle (see Figure 10-15). If your distance sensor is not centered, you can add a line in setup() that will center the servo. Example 10-6 shows the complete setup function with the servo centering line added.

Example 10-6. **The new setup function**

```
void setup()
{
  Serial.begin(9600);
  blinkNumber(8); // open port while flashing. Needed for Leonardo only

  lookBegin();
  moveBegin();
```

```
  moveSetSpeed(MIN_SPEED + 10) ; // Run at 10% above minimum speed
  softServoWrite(90, 2000);      // Add this line to center the servo
  Serial.println("Ready");
}
```

The call to softServoWrite centers the servo and waits for two seconds. If your sensor is not centered, follow these steps:

1. Switch the power off

2. Unscrew the servo shaft screw (see the instructions supplied with the Ping bracket)

3. Lift the Ping mounting bracket and reposition so it is facing forward

4. Replace the servo shaft screw

5. Power on and recheck

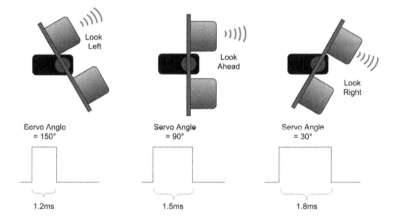

Figure 10-15. Servo used to scan left, center, and right

The servo angle is controlled by adjusting the pulse width on the Arduino pin connected to the servo. 1.5ms pulses will center the servo, and increasing or decreasing the pulse width will turn the servo one direction or the other.

The exact relationship between pulse width and servo angle varies across different servo products. If your servo turns right when it should turn left, swap the right and left servo angles in the servoAngles array:

```
// servo angles             left, right, center
const int servoAngles[] = {  150,   30,    90};
```

Arduino has a Servo library that can control up to 12 servos, however this is not used in this sketch for two reasons. The Servo library enables you to send an angle to the servo and carry on executing sketch code while the servo is being moved in the background, but your code must wait until the servo is facing the desired direction before requesting a reading from the distance sensor. However, the main reason not to use the Servo library is because it requires exclusive use of one of the Arduino chip's hardware timers (timer 1) and timers are in short supply on a standard Arduino chip (see Appendix F, *Arduino Pin and Timer Usage*).

The code to control the servo goes in a tab named Softservo (see Example 10-7).

Example 10-7. **The code from the Softservo tab**

```
/*******************************
 Softservo.ino
 software servo control without using timers
 note that these functions block until complete
 *******************************/

int servoPin;

void softServoAttach(int pin)
{
    servoPin = pin;
    pinMode(pin, OUTPUT);
}

// writes given angle to servo for given delay in milliseconds
void softServoWrite(int angle, long servoDelay)
{
    int pulsewidth = map(angle, 0, 180, 544, 2400); // width in microseconds
    do {
        digitalWrite(servoPin, HIGH);
        delayMicroseconds(pulsewidth);
        digitalWrite(servoPin, LOW);
        delay(20); // wait for 20 milliseconds
        servoDelay -= 20;
    } while(servoDelay >=0);
}
```

The softServoAttach function stores the pin number that the servo is attached to. The softServoWrite function converts the desired angle into a pulse width and creates the pulse using `digitalWrite` with a pulse width determined by a call to `delayMicroseconds`. The pulses are sent repeatedly for the duration of the given `servoDelay` which is a period sufficient for the servo to turn to the desired direction.

The `Look` code is similar to the code described at the beginning of this chapter, but here the `lookAt` function calls `softServoWrite` to rotate the servo instead of rotating the entire robot. Example 10-8 shows the `Look` tab used in the `myRobotScan` sketch.

Example 10-8. The modified Look tab code

```
/**********************
 code to look for obstacles
**********************/

// servo defines
const int sweepServoPin = 9;  // pin connected to servo
const int servoDelay    = 500; // time in ms for servo to move

const int   MIN_DISTANCE = 8;     // robot stops when object is nearer (in inches)
const int   CLEAR_DISTANCE = 24;  // distance in inches considered attracive to move
const int   MAX_DISTANCE = 150;   // the maximum range of the distance sensor

// servo angles            left, right, center
const int servoAngles[] = {  150,   30,    90};

const byte pingPin = 10; // digital pin 10

void lookBegin()
{
  irSensorBegin();     // initialize sensors
  softServoAttach(sweepServoPin);  /// attaches the servo pin to the servo object
}

// returns true if the given obstacle is detected
boolean lookForObstacle(int obstacle)
{
  switch(obstacle) {
     case  OBST_FRONT_EDGE: return irEdgeDetect(DIR_LEFT) && irEdgeDetect(DIR_RIGHT);
     case  OBST_LEFT_EDGE:  return irEdgeDetect(DIR_LEFT);
     case  OBST_RIGHT_EDGE: return irEdgeDetect(DIR_RIGHT);
     case  OBST_FRONT:      return  lookAt(servoAngles[DIR_CENTER]) <= MIN_DISTANCE;
  }
  return false;
}

// returns the distance of objects at the given angle
int lookAt(int angle)
{
  softServoWrite(angle, servoDelay ); // wait for servo to get into position
```

```
  int distance, samples;
  long cume;
  distance = samples = cume = 0;
  for(int i =0; i < 4; i++)
  {
    distance = pingGetDistance(pingPin);
    if(distance > 0)
    {
      //  printlnValue(" D= ",distance);
      samples++;
      cume+= distance;
    }
  }
  if(samples > 0)
    distance = cume / samples;
  else
    distance = 0;

  if( angle != servoAngles[DIR_CENTER])
  {
    Serial.print("looking at dir ");
    Serial.print(angle), Serial.print(" distance= ");
    Serial.println(distance);
    softServoWrite(servoAngles[DIR_CENTER], servoDelay/2);
  }
  return distance;
}


// function to check if robot can continue moving in current direction
// returns true if robot is not blocked moving in current direction
// this version only tests for obstacles in front
boolean checkMovement()
{
  boolean isClear = true; // default return value if no obstacles
  if(moveGetState() == MOV_FORWARD)
  {
    if(lookForObstacle(OBST_FRONT) == true)
    {
      isClear = false;
    }
  }
  return isClear;
}


// Look for and avoid obstacles using servo to scan
void roam()
{
  int distance = lookAt(servoAngles[DIR_CENTER]);
  if(distance == 0)
  {
    moveStop();
    Serial.println("No front sensor");
    return;  // no sensor
  }
```

```
  else if(distance <= MIN_DISTANCE)
  {
    moveStop();
    //Serial.print("Scanning:");
    int leftDistance  = lookAt(servoAngles[DIR_LEFT]);
    if(leftDistance > CLEAR_DISTANCE)  {
  //   Serial.print(" moving left: ");
      moveRotate(-90);
    }
    else {
      delay(500);
      int rightDistance = lookAt(servoAngles[DIR_RIGHT]);
      if(rightDistance > CLEAR_DISTANCE) {
      //  Serial.println(" moving right: ");
        moveRotate(90);
      }
      else {
       // Serial.print(" no clearence : ");
        distance = max( leftDistance, rightDistance);
        if(distance < CLEAR_DISTANCE/2) {
          timedMove(MOV_BACK, 1000); // back up for one second
          moveRotate(-180); // turn around
        }
        else {
          if(leftDistance > rightDistance)
            moveRotate(-90);
          else
            moveRotate(90);
        }
      }
    }
  }
}

// the following is based on loop code from myRobotEdge
// robot checks for edge and moves to avoid
void avoidEdge()
{
  if( lookForObstacle(OBST_FRONT_EDGE) == true)
  {
    Serial.println("left and right sensors detected edge");
    timedMove(MOV_BACK, 300);
    moveRotate(120);
    while(lookForObstacle(OBST_FRONT_EDGE) == true )
        moveStop(); // stop motors if still over cliff
  }
  else if(lookForObstacle(OBST_LEFT_EDGE) == true)
  {
    Serial.println("left sensor detected edge");
    timedMove(MOV_BACK, 100);
    moveRotate(30);
  }
  else if(lookForObstacle(OBST_RIGHT_EDGE) == true)
  {
```

```
        Serial.println("right sensor detected edge");
        timedMove(MOV_BACK, 100);
        moveRotate(-30);
    }
}
```

The `lookForObstacle` and `roam` functions are modified from the non-scanning version to use the appropriate servo angles for looking left, right, and center. The servo angles are stored in the array `servoAngle` (swap the left and right values if your servo turns in the wrong direction). The `lookAt` function now rotates the servo to the desired angle instead of moving the entire robot.

Remote Control 11

This chapter describes how to remotely control robot movement. Techniques for sending Serial commands as well as TV type infrared remote control are both explained. The example sketches enable you to command the robot to perform any of the higher level drive functions described in Chapter 7.

Hardware Required

- The TV remote control sketch requires an infrared decoder module. TSOP4838 (or the equivalent PNA4602) modules (Figure 11-1) have power and signal pins oriented to enable them to plug directly into the socket on the motor shield.

  You will also need an infrared remote control—almost any controller from a TV or DVD player will do.

Figure 11-1. *Infrared Decoder Module*

Sketches Used in This Chapter

- myRobotSerialRemote.ino—enables the robot to be controlled by commands from the serial port.
- myRobotRemote.ino - uses commands from a TV type remote to control the robot.

Figure 11-2 shows the modules used in this chapter.

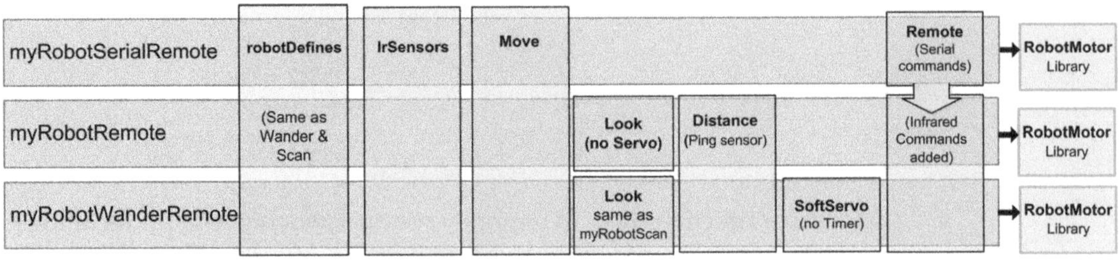

myRobotSerialRemote	robotDefines	IrSensors	Move					Remote (Serial commands)		RobotMotor Library
myRobotRemote	(Same as Wander & Scan			Look (no Servo)	Distance (Ping sensor)			(Infrared Commands added)		RobotMotor Library
myRobotWanderRemote				Look same as myRobotScan			SoftServo (no Timer)			RobotMotor Library

Figure 11-2. *Remote Control sketches*

Design of the Remote Control Code

The code to handle remote control functions is contained in a module named Remote that appears as a tab in sketches introduced in this chapter. This module:

- Defines constants that identify each command.
- Matches received data to a command.
- Executes a function associated with each command which activates the appropriate action.

Here are the commands used in the remote control example:

```
const char MOVE_FORWARD = 'f'; // move forward
const char MOVE_BACK    = 'b'; // move backward
const char PIVOT_CCW    = 'C'; // rotate 90 degrees CCW
const char PIVOT_CW     = 'c'; // rotate 90 degrees CW
const char PIVOT        = 'p'; // rotation angle (minus rotates CCW)
const char HALT         = 'h'; // stop moving
```

These constants are used to switch program execution to a function associated with each command:

```
void processCommand(int cmd, int val)
{
  switch(cmd)
  {
```

```
        case MOVE_FORWARD : changeCmdState(MOV_FORWARD); moveForward();   break;
        case MOVE_BACK    : changeCmdState(MOV_BACK);    moveBackward();  break;
        case PIVOT_CCW    : changeCmdState(MOV_ROTATE);  moveRotate(-90); break;
        case PIVOT_CW     : changeCmdState(MOV_ROTATE);  moveRotate(90);  break;
        case PIVOT        : changeCmdState(MOV_ROTATE);  moveRotate(val); break;
        case HALT         : changeCmdState(MOV_STOP);    moveStop();      break;
    }
}
```

Before calling a movement function, a function named changeCmdState is called to store the current command state. This enables the robot logic to be aware of what it was last asked to do so it can make decisions if it encounters obstacles while trying to execute the commanded movement.

To get this code working with serial commands, all that is needed is to add a function that passes serial data to the processCommand function. Example 11-1 shows the code for the Remote tab that supports simple serial remote control.

Example 11-1. **Remote tab code for simple serial remote control**

```
// robot remote commands
// This version is for serial commands

// Command constants

const char MOVE_FORWARD = 'f'; // move forward
const char MOVE_BACK    = 'b'; // move backward
const char MOVE_LEFT    = 'l'; // move left
const char MOVE_RIGHT   = 'r'; // move right
const char PIVOT_CCW    = 'C'; // rotate 90 degrees CCW
const char PIVOT_CW     = 'c'; // rotate 90 degrees CW
const char PIVOT        = 'p'; // rotation angle (minus rotates CCW)
const char HALT         = 'h'; // stop moving

// not used in this example
const char MOVE_SPEED   = 's';
const char MOVE_SLOWER  = 'v'; // reduce speed
const char MOVE_FASTER  = '^'; // increase speed

int commandState = MOV_STOP;    // what robot is told to do

void remoteService()
{
  if(Serial.available() )
  {
    int cmd = Serial.read();
    processCommand(cmd);
  }
}


void changeCmdState(int newState)
```

```
{
  if(newState != commandState)
  {
    Serial.print("Changing Cmd state from "); Serial.print( states[commandState]);
    Serial.print(" to "); Serial.println(states[newState]);
    commandState = newState;
  }
}

void processCommand(int cmd)
{
  int val = 0;
  if( cmd == PIVOT || cmd == SPEED) {
    val =  Serial.parseInt();
  }
  processCommand(cmd, val);
}

void processCommand(int cmd, int val)
{
  byte speed;
  Serial.write(cmd); // echo
  switch(cmd)
  {
    case MOVE_LEFT    : changeCmdState(MOV_LEFT);     moveLeft();       break;
    case MOVE_RIGHT   : changeCmdState(MOV_RIGHT);    moveRight();      break;
    case MOVE_FORWARD : changeCmdState(MOV_FORWARD);  moveForward();    break;
    case MOVE_BACK    : changeCmdState(MOV_BACK);     moveBackward();   break;
    case PIVOT_CCW    : changeCmdState(MOV_ROTATE);   moveRotate(-90);  break;
    case PIVOT_CW     : changeCmdState(MOV_ROTATE);   moveRotate(90);   break;
    case PIVOT        : changeCmdState(MOV_ROTATE);   moveRotate(val);  break;
    case HALT         : changeCmdState(MOV_STOP);     moveStop();       break;
    case SPEED        : speed = val;             moveSetSpeed(speed);   break;
  }
}
```

This code adds a function named remoteService that is called from the main sketch to check if any remote commands have been received. The remoteSer vice function will be exanded later in this chapter to support other remote control inputs.

You may have noticed that there are two functions named processCommand. The one that takes a single parameter tests if a second parameter is required (as in the case of the PIVOT command) and if so gets this using the Serial Stream parseInt function.

Example 11-2 shows the main sketch code from the example, myRobotSerial-Remote that responds to the serial commands.

Example 11-2. **Main sketch code**

```
/****************************************************************************
myRobotSerialRemote.ino

Robot sketch with serial remote commands

Created by Michael Margolis 10 June 2012
****************************************************************************/

#include <AFMotor.h>   // adafruit motor shield library
#include "RobotMotor.h"    // 2wd or 4wd motor library

#include "robotDefines.h"  // global defines

// Setup runs at startup and is used configure pins and init system variables
void setup()
{
  Serial.begin(9600);
  while(!Serial);  // only needed for leonardo

  moveBegin();
  moveSetSpeed(MIN_SPEED + 10) ;  // Run at 10% above minimum speed
}

void loop()
{
  remoteService(); // wait for serial commands
}

// function to check if robot can continue moving when taking evasive action
// returns true if robot is not blocked when moving to avoid obstacles
// this 'placeholder' version always returns true
boolean checkMovement()
{
  return true;
}
```

If you have a wireless device that passes serial data such as a Bluetooth module, you can wirelessly control the robot by connecting the serial output of the adapter to the Arduino serial input and wiring up the power leads. If you are using a Leonardo, note that the TX/RX pins (digital 1 and 0) are accessed through Serial1 rather than Serial, so modify your code accordingly (you'll need to replace all instances of Serial with Serial1 in all the tabs of your sketch).

Controlling the Robot with a TV Type IR Remote

The remote code can be expanded to support the decoding of IR remote controls. To do this, an IR module is used to receive and condition the IR pulses so they can be decoded by Arduino.

Installing the IR Decoder Chip

The IR receiver module looks something like a three pin transistor with a bulge for the IR sensor lens. The module is plugged into the shield as shown in Figure 11-3 and Figure 11-4. It is polarized, so make sure it is facing the direction shown or you can damage the module.

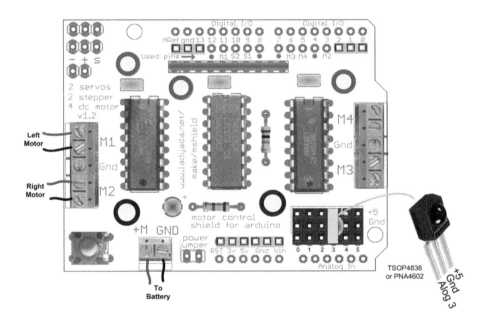

Figure 11-3. *IR Receiver Module*

Figure 11-4. IR Receiver Module plugged into the motor shield

If the receiver module does not plug securely into the socket, use a long-nose pliers to twist the ends of each of the three leads 90 degrees, as shown in Figure 11-5.

Figure 11-5. IR Receiver Module with leads twisted to for better fit into socket

The IR Remote Software

The low level decoding of the infrared signal is handled by an Arduino library named `IRremote` that is included with the book's download code.

If you need help installing a library, see "Installing Third-Party Libraries" (page 83).

You don't need to understand how the library works in order to use it, but if you are curious, the following is an overview of how the library works.

The `IRremote` library uses an `irrecv` object to decode the pulses from the IR Receiver.

The IRremote Library and Pin Assignments

The code to create the `irrecv` object is:

```
IRrecv irrecv(irReceivePin);
```

The `irReceivePin` is the pin that the module is connected to. This pin is defined at the top of the main sketch tab:

```
const byte irReceivePin = A3;
// analog pin 3
```

The Arduino analog input pins can also be used as digital pins, but the pin numbers are not the same—analog pin 3 is not digital pin 3! The A3 constant is the Arduino way of referring to the digital pin number associated with the analog input (the `irrecv` object expects the digital pin number). On Leonardo, analog input 3 is used as digital input 21; on a standard AT-mega328 board like the Uno, A3 is digital pin 17. If you use Arduino constants to refer to the digital pin assignments for the analog input pins, the correct values will automatically be assigned.

A numeric value is provided for each remote keypress detected. The specific key values decoded will depend on the remote controller you use. Example 11-3 shows code for the `Remote` tab with support for the IR receiver.

Example 11-3. **The Remote tab code**

```
// robot remote commands

#include <IRremote.h>              // IR remote control library

IRrecv irrecv(irReceivePin);

decode_results results;

// Command constants

const char MOVE_FORWARD = 'f'; // move forward
const char MOVE_BACK    = 'b'; // move backward
const char MOVE_LEFT    = 'l'; // move left
const char MOVE_RIGHT   = 'r'; // move right
const char PIVOT_CCW    = 'C'; // rotate 90 degrees CCW
const char PIVOT_CW     = 'c'; // rotate 90 degrees CW
```

```
const char PIVOT         = 'p'; // rotation angle (minus rotates CCW)
const char HALT          = 'h'; // stop

// not used in this example
const char MOVE_SPEED     = 's';
const char MOVE_SLOWER    = 'v'; // reduce speed
const char MOVE_FASTER    = '^'; // increase speed

//IR remote keycodes:replace this with codes for your remote
// See text for procedure for obtaining codes.
const long IR_MOVE_FORWARD = 1064;
const long IR_MOVE_BACK    = 3112;
const long IR_MOVE_LEFT    = 1128;
const long IR_MOVE_RIGHT   = 2152;
const long IR_PIVOT_CW     = 136;
const long IR_PIVOT_CCW    = 1160;
const long IR_HALT         = 2216;

int commandState = MOV_STOP;    // what robot is told to do

void remoteBegin(byte irPin)
{
  irrecv.enableIRIn(); // Start the receiver
}

void remoteService()
{
  if (irrecv.decode(&results))
  {
    if (results.decode_type != UNKNOWN)
    {
      //Serial.println(results.value); // uncomment to see raw result
      convertIrToCommand(results.value);
    }
    irrecv.resume(); // Receive the next value
  }
  // additional support for serial commands
  if(Serial.available() )
  {
    int cmd = Serial.read();
    processCommand(cmd);
  }
}

void convertIrToCommand(long value)
{
  {
  switch(value)
  {
    case  IR_MOVE_LEFT    : processCommand(MOVE_LEFT);    break;
    case  IR_MOVE_RIGHT   : processCommand(MOVE_RIGHT);   break;
    case  IR_MOVE_FORWARD : processCommand(MOVE_FORWARD); break;
    case  IR_MOVE_BACK    : processCommand(MOVE_BACK);    break;
    case  IR_PIVOT_CCW    : processCommand(PIVOT_CCW);    break;
```

```
      case  IR_PIVOT_CW    :  processCommand(PIVOT_CW);      break;
      case  IR_HALT        :  processCommand(HALT);          break;
//    case  IR_SLOWER      :  processCommand(SLOWER);        break;
//    case  IR_FASTER      :  processCommand(FASTER);        break;
    }
  }
}

void changeCmdState(int newState)
{
  if(newState != commandState)
  {
    Serial.print("Changing Cmd state from "); Serial.print( states[commandState]);
    Serial.print(" to "); Serial.println(states[newState]);
    commandState = newState;
  }
}

void processCommand(int cmd)
{
  int val = 0;
  if( cmd == MOVE_SPEED) {
    val =  Serial.parseInt();
  }
  else if( cmd == PIVOT) {
    val =  Serial.parseInt();
  }
  processCommand(cmd, val);
}

void processCommand(int cmd, int val)
{
  byte speed;
  //Serial.write(cmd); // uncomment to echo
  switch(cmd)
  {
    case MOVE_LEFT    : changeCmdState(MOV_LEFT);     moveLeft();        break;
    case MOVE_RIGHT   : changeCmdState(MOV_RIGHT);    moveRight();       break;
    case MOVE_FORWARD : changeCmdState(MOV_FORWARD);  moveForward();     break;
    case MOVE_BACK    : changeCmdState(MOV_BACK);     moveBackward();    break;
    case PIVOT_CCW    : changeCmdState(MOV_ROTATE);   moveRotate(-90);   break;
    case PIVOT_CW     : changeCmdState(MOV_ROTATE);   moveRotate(90);    break;
    case PIVOT        : changeCmdState(MOV_ROTATE);   moveRotate(val);   break;
    case HALT         : changeCmdState(MOV_STOP);     moveStop();        break;
    case MOVE_SPEED   : speed = val;      moveSetSpeed(speed);           break;
//    case SLOWER      : moveSlower(speedIncrement);                     break;
//    case FASTER      : moveFaster(speedIncrement);                     break;
    case '\r' : case '\n': break; // ignore cr and lf
    default :  Serial.print('['); Serial.write(cmd); Serial.println("] Ignored");  break;
  }
}
```

Timers and IRremote Library

The standard IRremote library uses Timer 2 and, at the time of writing, did not support the Leonardo board. However, Timer 1 is the only available timer when using the 4WD with the Leonardo board. The version of this library included with the example code (see "How to Contact Us" (page xv)) supports the Leonardo board and is modified to use Timer 1.

"Pin and Timer Tables" (page 237) shows how the sketches in this book use the timers. These timers were selected so the same code can be used with the 2 motor and 4 motor robots with either the Arduino Uno or Arduino Leonardo boards. If you want to use a timer other than Timer 1 and you are sure that another timer is free, then the information on "Modifying a Library to Change Timer Allocation" (page 236) will help you modify the library for use with a different Timer.

To use this code with your remote, you need to replace the IR commands with the ones your remote controller sends. Figure 11-6 shows a typical controller with a suggested key assignment but you can choose any keys you want.

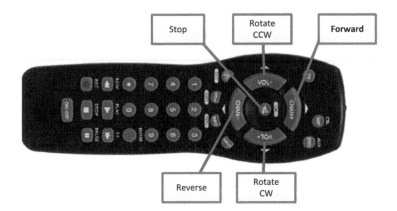

Figure 11-6. *Remote Controller Command Buttons*

You can use the sketch shown in Example 11-4 to display the actual IR codes that are sent. After you upload and run the sketch, it will prompt you to press a key for each command to be learned. These are: forward, reverse, left, right, pivot counterclockwise, pivot clockwise, and stop. The decoded value will be displayed for each recognized keypress. After all the keys are learned, the codes are written to the Serial Monitor in a format that you can copy into the Re mote tab below the comment line that reads: //IR remote keycodes:replace this with codes for your remote.

Example 11-4. **Learning remote sketch**
```
/*
 * LearningRemote.cpp
```

```
*/

#include <IRremote.h>        // IR remote control library

const int irPin = A3;    // analog input pin 3 (digital 17)
const long NO_KEY = -1;
const long TIMEOUT = 5000; //max number of milliseconds to wait for a key (5 secs)
const int KEYCOUNT = 7; // the number of key codes supported

long  irKeyCodes[ KEYCOUNT]; // this will store raw codes for all keys
char * remoteKeyNames[KEYCOUNT] =
  {"Forward", "Back", "Left",  "Right", "PivotCW", "PivotCCW", "Halt" };
// not used: Slower, Faster

IRrecv irrecv(irPin);       // create the IR receive object
decode_results results;    // ir data goes here

void setup()
{
  Serial.begin(9600);
  while(!Serial);   // only needed for leonardo

  irrecv.enableIRIn(); // Start the ir receiver
  learnKeycodes();
  printConstants();
  Serial.println();
  Serial.println("Now press the remote keys to verify correct detection");
}

void loop()
{
  long key = getIrKeycode(TIMEOUT);
  if( key!= NO_KEY)
  {
    int index =  findKey(key);
    if( index != NO_KEY)
    {
      Serial.println(remoteKeyNames[index]);
    }
  }
}

// get remote control codes
// the key map should be set to zero before calling this
void learnKeycodes()
{
  Serial.println("Ready to learn remote codes");
  for(int i = 0; i < KEYCOUNT;  )
  {
    //delay(100);
    Serial.println();
    Serial.print("press remote key for ");
    Serial.print( remoteKeyNames[i]);
    long key = getIrKeycode(TIMEOUT);
```

```
      if( key > 0 )
      {
          Serial.println(", release key ...");
          long temp;
          do {
              temp = getIrKeycode(200);
          }
          while( temp == key);
          if( findKey(key) == NO_KEY)
          {
            Serial.print(" -> using code ");
            Serial.print( key );
            Serial.print(" for ");
            Serial.println(remoteKeyNames[i]);
            irKeyCodes[i] = key;
            i++;
          }
          else
          {
              Serial.println("key already assigned");
          }
      }
      else continue;
  }
  Serial.println("Done\n");
}


// wait up to timeout milliseconds for a key
long getIrKeycode(long timeout)
{

  flushKeys();

  long key = NO_KEY;
  unsigned long startTime = millis();
  while( millis() - startTime < timeout )
  {
    if( irrecv.decode(&results) ) {
      key = results.value;
      //Serial.println(key, HEX);
      irrecv.resume(); // Receive the next value
      if(key != NO_KEY ) {
        break;
      }
    }
  }
  return key;
}

//clear the buffer
void flushKeys()
{
  while(irrecv.decode(&results))
```

```
    irrecv.resume();
  results.value = -1;
}

// returns the index for the given key code if found
// returns NO_KEY if code is not found
int findKey(long code)
{
  for(int i=0; i < KEYCOUNT; i++ )
  {
    if(irKeyCodes[i] == code)
       return i;
  }
  return NO_KEY;
}

void printConstants()
{
 int i = 0;
 Serial.println("//IR remote keycodes:");
 Serial.print("const long IR_MOVE_FORWARD = "); Serial.print(irKeyCodes[i++]);
       Serial.println(";");
 Serial.print("const long IR_MOVE_BACK   = "); Serial.print(irKeyCodes[i++]);
       Serial.println(";");
 Serial.print("const long IR_MOVE_LEFT   = "); Serial.print(irKeyCodes[i++]);
       Serial.println(";");
 Serial.print("const long IR_MOVE_RIGHT  = "); Serial.print(irKeyCodes[i++]);
       Serial.println(";");
 Serial.print("const long IR_PIVOT_CW    = "); Serial.print(irKeyCodes[i++]);
       Serial.println(";");
 Serial.print("const long IR_PIVOT_CCW   = "); Serial.print(irKeyCodes[i++]);
       Serial.println(";");
 Serial.print("const long IR_HALT        = "); Serial.print(irKeyCodes[i++]);
       Serial.println(";");

 Serial.println(); Serial.println("Copy the above lines to the Remote tab");
}
```

The IR remote example sketch shown in Example 11-5 is similar to the example earlier in this chapter with some additional lines for the IR remote.

Example 11-5. *IR remote sketch code*
```
/*****************************************************************************
myRobotRemote.ino

Robot sketch with remote commands
Note: if motors don't turn, check that irRemoteInt.h uses Timer1, not Timer2

Michael Margolis 28 May 2012
```

```
**************************************************************************/

#include <AFMotor.h>   // adafruit motor shield library
#include "RobotMotor.h"    // 2wd or 4wd motor library

#include "robotDefines.h"  // global defines

const byte irReceivePin = A3; /// analog pin 3

// Setup runs at startup and is used configure pins and init system variables
void setup()
{
  Serial.begin(9600);
  blinkNumber(8); // open port while flashing. Needed for Leonardo only

  lookBegin();
  moveBegin();
  remoteBegin(irReceivePin);  /// added Remote tab

  moveSetSpeed(MIN_SPEED + 10) ;  // Run at 10% above minimum speed
  Serial.println("Ready");
}

void loop()
{
  remoteService();
}

// function to indicate numbers by flashing the built-in LED
void blinkNumber( byte number) {
    pinMode(LED_PIN, OUTPUT); // enable the LED pin for output
    while(number--) {
      digitalWrite(LED_PIN, HIGH); delay(100);
      digitalWrite(LED_PIN, LOW);  delay(400);
    }
}
```

The remoteBegin(irReceivePin); function called in setup initializes the IR-decode library. ireReceivePin is the pin the ir decoder module is connected to, in this case, analog pin 3. Because the library expects a digital pin number, the Arduino constant A3 is used.

Enhancing Your Robot Ɐ

Experienced software engineers use a variety of techniques to help manage complex projects. This section provides a list of useful tips for designing and building complex robotics projects.

Planning

Think Before You Code

It helps to think about your project and be clear on what you want it to achieve before you start coding. Tinkering around without a plan is a good way to learn and to have fun, but it can make a large project too cumbersome to manage.

Avoid Feature Bloat

Don't add features that you don't need. Features added 'just in case' may seem like a good way to save time later on, but nine times out of ten, the features you actually need will require changes to the code you wrote that are usually more troublesome than writing the feature from scratch after you are clear on what you really need.

Don't Reinvent the Wheel

See what's out there that you can reuse. The Arduino community has made available a vast collection of useful software. Even if you don't end up using something off the shelf, seeing how others have solved problems similar to yours can inform your own solutions.

Structure to Reflect Functionality

Think about functional associations when organizing your code. Grouping similar functionality together enables you to create modules which can be tested in isolation from the rest of your code, simplifying debugging and reducing the likelihood that adding functionality later will have side effects on other parts of your sketch.

Use Clear Names for Functions and Variables

Each function should have a single clear purpose. Choose a name that reflects that purpose so when you later need to debug, add or change functionality, it will be clear what each function or variable is doing. A few extra seconds spent finding evocative names when you are coding can save hours later on when you are trying to figure out what a piece of code is intended to do.

Implementing a Complex Project

Test Often

Testing after each major addition or change in code saves time because you will find and fix problems more quickly. It may seem more efficient to wait until you have lots of code completed before stopping to test, but when the inevitable bugs arise, you can waste a huge amount of time just trying to locate which part of the code is causing the problem.

Simplify

Spending time simplifying code will be repaid in reduced debug time. Complex code can be difficult to debug or enhance, particularly when you come back to it after a while. Looking at each completed function with an eye to seeing if there is a simpler way of achieving the functionality can result in cleaner code that is easier to maintain.

If It Is Awkward, Start Over

Don't be afraid to throw away prototype code that becomes a burden. Sometimes you need to tinker to get things working, but if this ends up making your code a tangled mess, use what you have learned to rethink your structure and start over.

Don't Confuse Activity with Progress

If you are not making progress, stop and take a break and come back fresh. It is easier taking in the big picture after a break, particularly if you pause to get a clear picture in your mind of the problem you are trying to solve and a list of the assumptions you are making about what is standing in your way.

Experiment

If what you have tried isn't working, try something new. Software problems may actually be a hardware issue (and vice versa).

Be Tenacious

Interesting projects usually come with difficult problems—overcoming these is part of the reward for a job well done.

Have Fun

Isn't that why you started this project in the first place?

Using Other Hardware with Your Robot

B

You may want to add more capability to your robot or perhaps substitute different hardware than the items covered in the text. This chapter describes how to use some common alternative components.

Alternative Motor Controllers

Ardumoto

This popular H-bridge can be used instead of the Adafruit shield described in the text if you have a two wheeled robot (the shield only supports two motors). It also lacks the convenient layout for the analog sensors and you will need to add two 3 pin headers for the servo and distance sensor connections. The Motor code for Ardumoto is shown in Example B-1.

Continuous Rotation Servos

Continuous rotation servos are hobby servos modified to rotate continuously with a speed and rotation direction controlled by the Servo library that comes with Arduino. The servo rotates in one direction as the angle written to the servo is increased from 90 degrees; it rotates in the other direction when the angle is decreased from 90 degrees. The actual direction forward or backward depends on how you have the servos attached. Continuous rotation servos may not stop rotating when writing exactly 90 degrees. Some servos have a small potentiometer you can trim to adjust for this, or you can add or subtract a few degrees to the `motorStopAngle` element to stop the servo. This version uses digital pins 7 and 8 but you can change this by altering the elements of the `servoPins` array (the first element for all the arrays is the left servo, the second is the right servo). See Example B-2.

These functions convert requests to set the motor speed into servo angles that are written to the continuous rotation servos. The conversion is performed using the Arduino map function.

This code uses the Servo library. If you want to build the infrared remote control project with continuous rotation servos, you will need to ensure that the `IRremote` library is configured to use a timer other than Timer 1) because the Servo library requires Timer 1. See **"Modifying a Library to Change Timer Allocation" (page 236)** *for timer usage and details on how to configure timers for the `IRremote` library.*

Example B-1. **RobotMotor library code for the Ardumoto shield**

```
/***********************************************************
    RobotMotor.cpp // Ardumoto version
    low level motor driver for use with ardumoto motor shield and 2WD robot

    Michael Margolis May 8 2012
***********************************************************/

#include <Arduino.h>
#include "RobotMotor.h"

const int differential = 0; // % faster left motor turns compared to right

/****** motor pin defines *************/
// Pins connected to the motor driver. The PWM pins control the speed, and the
//   other pins are select forward and reverse

// Motor uses pins : 3,11,12,13
const byte M_PWM_PIN[2]  = {11,3};  // ardumoto v13
const byte M_DIR_PIN[2]  = {13,12};
/* end of motor pin defines */

int  motorSpeed[2]  = {0,0}; // motor speed stored here (0-100%)

// tables hold time in ms to rotate robot 360 degrees at various speeds
// this enables conversion of rotation angle into timed motor movement
// The speeds are percent of max speed
// Note: low cost motors do not have enough torque at low speeds so
// the robot will not move below this value
// Interpolation is used to get a time for any speed from MIN_SPEED to 100%

const int MIN_SPEED = 40; // first table entry is 40% speed
const int SPEED_TABLE_INTERVAL = 10; // each table entry is 10% faster speed
const int NBR_SPEEDS =  1 + (100 - MIN_SPEED)/ SPEED_TABLE_INTERVAL;

int speedTable[NBR_SPEEDS]  = {40,    50,   60,   70,   80,   90,  100}; // speeds
int rotationTime[NBR_SPEEDS] = {5500, 3300, 2400, 2000, 1750, 1550, 1150}; // time
```

```
void motorBegin(int motor)
{
  pinMode(M_DIR_PIN[motor], OUTPUT);
  motorStop(motor);
}

// speed range is 0 to 100
void motorSetSpeed(int motor, int speed)
{
    motorSpeed[motor] = speed;           // save the value
    speed = map(speed, 0,100, 0,255);    // scale to PWM range
    analogWrite(M_PWM_PIN[motor], speed);  // write the value
}

void motorForward(int motor, int speed)
{
  digitalWrite(M_DIR_PIN[motor], HIGH);
  motorSetSpeed(motor, speed);
}

void motorReverse(int motor, int speed)
{
  digitalWrite(M_DIR_PIN[motor], LOW);
  motorSetSpeed(motor, speed);
}

void motorStop(int motor)
{
  analogWrite(M_PWM_PIN[motor], 0);
}

void motorBrake(int motor)
{
  // Ardumoto does not support brake, so just stop the motor
  analogWrite(M_PWM_PIN[motor], 0);
}
```

Example B-2. **RobotMotor library header for continuous rotation servos**

```
/*******************************************************
    RobotMotor.cpp // continuous rotation servo version
    low level motor driver for use with continuous rotation servos and 2WD robot

    Copyright Michael Margolis May 8 2012
*******************************************************/

#include <Arduino.h>
#include <Servo.h>
#include "RobotMotor.h"

Servo myservo[2];               // create instances for two servos
```

```
const int MAX_ANGLE    = 60;     // number of degrees that motor driven at max speed
const int servoPins[2] = {7,8}; // digital pins connected to servos:(left,right)

                                // change sign to reverse direction of the motor
int motorSense[2] = {1,-1};      // 1 increases angle for forward, -1 decreaes

int motorStopAngle[2] = {90,90}; // inc or dec so motor stops when motorStop is called

int motorSpeed[2] = {0,0};       // left and right motor speeds stored here (0-100%)

// tables hold time in ms to rotate robot 360 degrees at various speeds
// this enables conversion of rotation angle into timed motor movement
// The speeds are percent of max speed
// Note: low cost motors do not have enough torque at low speeds so
// the robot will not move below this value
// Interpolation is used to get a time for any speed from MIN_SPEED to 100%

const int MIN_SPEED = 40; // first table entry is 40% speed
const int SPEED_TABLE_INTERVAL = 10; // each table entry is 10% faster speed
const int NBR_SPEEDS =  1 + (100 - MIN_SPEED)/ SPEED_TABLE_INTERVAL;

int speedTable[NBR_SPEEDS]   = {40,    50,   60,   70,   80,   90,  100}; // speeds
int rotationTime[NBR_SPEEDS] = {5500, 3300, 2400, 2000, 1750, 1550, 1150}; // time

void motorBegin(int motor)
{
    myservo[motor].attach(servoPins[motor]);
}

// speed range is 0 to 100
void motorSetSpeed(int motor, int speed)
{
    motorSpeed[motor] = speed;          // save the value
}

void motorForward(int motor, int speed)
{
  motorSetSpeed(motor, speed);
  int stopAngle = motorStopAngle[motor];
  int maxSpeedAngle =  stopAngle + (MAX_ANGLE *  motorSense[motor]);
  int angle = map(speed, 0,100, stopAngle, maxSpeedAngle);
  myservo[motor].write(angle);
}

void motorReverse(int motor, int speed)
{
  motorSetSpeed(motor, speed);
  int stopAngle = motorStopAngle[motor];
  int maxSpeedAngle =  stopAngle - (MAX_ANGLE *  motorSense[motor]);
  int angle = map(speed, 0,100, stopAngle, maxSpeedAngle);
  myservo[motor].write(angle);
}
```

```
void motorStop(int motor)
{
  myservo[motor].write(motorStopAngle[motor]);
}

void motorBrake(int motor)
{
  myservo[motor].write(motorStopAngle[motor]);
}
```

Debugging Your Robot | C

Complex projects inevitably throw up obstacles in the form of bugs. As these arise, you can congratulate yourself for choosing such a challenging project and bear in mind the satisfaction you will feel when all the problems have been overcome. Here is some software that should help you find and fix problems you may encounter.

Identify the Symptoms and Localize the problem

Seeing What the Robot Is Doing

Visualizing data from the sensors in real time can be tremendous help in understanding what is actually happening in your sketch. Figure C-1 shows the screen from a Processing sketch that enables you to easily view Arduino values.

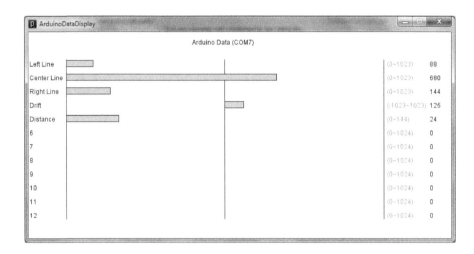

Figure C-1. *Arduino Data Displayed in Processing*

Figure C-1 depicts the analogRead values from left, center and right and sensors used in line detection. The grey numbers in parentheses on the right indicate the possible range of values, the following number is the numeric value sent from Arduino.

The figure shows the values when the robot is straying slightly to the right of a dark line it is trying to follow (the values increase when the sensor is over the line). The Position value goes positive when straying right and negative when left. The Position value is used in the line following sketch to adjust the robot direction so it stays on the line. The Distance value is in inches and is obtained from the ping distance sensor.

The Processing sketch expects data in the following format; fields are separated by commas:

- The string: Data
- The row to display; the first row is row 1
- The value to display
- The newline character "\n"

For example, sending "Data,2,680\n" will display a bar with a value of 680 on the second line.

You can send labels for each line by sending a string such as: "Label,2,Center Line\n", which will tag the second row with the label "Center Line".

The data range can be sent using a string such as "Range,5,0,144\n" which will set the range of the fifth line from 0 to 144.

The easiest way to send this data is to add the DataDisplay tab to your sketch and call the functions to format and send the data:

- `sendData(row, value);` sends the specified value for display on the given row
- `sendLabel(row, label);` sends the specified label for display on the given row
- `sendRange(row, minimum, maximum);` sends the minimum and maximum values for the specified row

The sketch named `myRobotDebug` contains the DataDisplay tab and provides an example of how to send data to the Processing. Example C-1 shows the main sketch code.

Example C-1. **myRobotDebug main sketch code**

```
/***********************************************************
MyRobotDebug.ino

Simple debug example using Processing debug monitor
This version displays values from the line and distance sensors

***********************************************************/
#include "robotDefines.h"  // global defines

const byte pingPin = 10; // Distance sensor connected to digital pin 10

enum {DATA_start, DATA_LEFT, DATA_CENTER, DATA_RIGHT,DATA_DRIFT,DATA_DISTANCE,
      DATA_nbrItems};

char* labels[]= {"","Left Line","Center Line","Right Line","Drift","Distance"};
int minRange[]= { 0,          0,           0,          0, -1023,        0 };
int maxRange[]= { 0,       1023,        1023,       1023,  1023,      144};

// Setup runs at startup and is used configure pins and init system variables
void setup()
{
  Serial.begin(9600);
  while(!Serial);  // only needed for leonardo

  dataDisplayBegin(DATA_nbrItems, labels, minRange, maxRange );
  Serial.println("Ready");
}

void loop()
{
  lineSense();
  int distance = pingGetDistance(pingPin);
  sendData(DATA_DISTANCE, distance); // send distance
```

```
}

/****************************
   Line   Sensor code
****************************/
// defines for locations of sensors
const int SENSE_LINE_LEFT  = 0;
const int SENSE_LINE_RIGHT = 1;
const int SENSE_LINE_CENTER = 2;

//returns drift - 0 if over line, minus value if left, plus if right
int lineSense()
{
   int leftVal = analogRead(SENSE_LINE_LEFT);
   int centerVal = analogRead(SENSE_LINE_CENTER);
   int rightVal = analogRead(SENSE_LINE_RIGHT);

   sendData(DATA_LEFT, leftVal);       // send left sensor value
   sendData(DATA_CENTER, centerVal);   // send center sensor value
   sendData(DATA_RIGHT, rightVal);     // send right sensor values

   int leftSense = centerVal - leftVal;
   int rightSense = rightVal - centerVal;
   int drift = leftVal- rightVal   ;

   sendData(DATA_DRIFT, drift);      // send drift sensor values

   return drift;
}

// Returns the distance in inches
// this will return 0 if no ping sensor is connected or
// the distance is greater than around 10 feet
int pingGetDistance(int pingPin)
{
   long duration, cm;

   pinMode(pingPin, OUTPUT);
   digitalWrite(pingPin, LOW);
   delayMicroseconds(2);
   digitalWrite(pingPin, HIGH);
   delayMicroseconds(5);
   digitalWrite(pingPin, LOW);

   pinMode(pingPin, INPUT);
   duration = pulseIn(pingPin, HIGH, 20000); // if a pulse does not arrive in 20 ms then
                                             // the ping sensor is not connected
   if(duration >=20000)
      return 0;
```

```
  // convert the time into a distance
  cm = duration / 29 / 2;
  return (cm * 10) / 25 ; // convert cm to inches
}
```

The Arduino code that sends the data is in the tab named DataDisplay
(Example C-2), you can copy the code into any sketch you want to debug, or
you can simply add the tab to the sketch.

Example C-2. **the DataDisplay tab code**

```
// DataDisplay

void dataDisplayBegin(int nbrItems, char* labels[], int minRange[], int maxRange[] )
{
    for(int i = 1; i < nbrItems; i++)
    {
      sendLabel(i, labels[i]);
      sendRange(i,  minRange[i], maxRange[i]);
    }
}

void sendLabel( int row, char *label)
{
   sendString("Label"); sendValue(row); sendString(label); Serial.println();
}

void sendRange( int row, int min, int max)
{
   sendString("Range"); sendValue(row); sendValue(min); sendValue(max); Serial.println();
}


void sendData(int row, int val)
{
    sendString("Data"); sendValue(row); sendValue(val); Serial.println();
}

void sendValue( int value)
{
    Serial.print(value); Serial.print(",");
}

void sendString(char *string)
{
    Serial.print(string); Serial.print(",");
}
```

The Processing sketch is called `ArduinoDataDisplay` and is located in the `Pro` `cessing` folder of the example code download (see "How to Contact Us" (page xv) for the download location). Example C-3 shows the code.

Example C-3. **Processing sketch for displaying data**

```
/*
 * ArduinoDataDisplay
 * based on Arduino Cookbook code from Recipe 4.4
 *
 * Displays bar graphs of sensor data sent as CSV from Arduino
 * in all cases, N is the Row to be associated with the given message
 * Labels sent as: "Label,N,the label\n"  // "the label" is used for Row N
 * Range sent as : "Range,N,Min, Max\n"  // Row N has a range from min to max
 *     if Min is negative then the bar grows from the midpoint of Min and Max,
 *     else the bar grows from Min
 * Data sent as:  "Data,N,val\n"   // val is plotted for row N
 */

short portIndex = 1;  // select the com port, 0 is the first port

int maxNumberOfRows = 12;
int graphWidth     = 600;
int displayWidth   = 1024;
int displayHeight  = 800;

int fontSize = 12;
PFont fontA;

int windowWidth;
int windowHeight;

int graphHeight;
int rectCenter;
int rectLeft;
int rectRight;
int topMargin;
int bottomMargin;
int leftMargin = 50;
int rightMargin = 80;

int textHeight;

ArrayList<String> labelList = new ArrayList<String>();
int [] values    = { 0, 0, 0, 0, 0, 0, 0, 0, 0, 0, 0, 0};
int [] rangeMin = { 0, 0, 0, 0, 0, 0, 0, 0, 0, 0, 0, 0};
int [] rangeMax = { 0, 1024, 1024, 1024, 1024, 1024, 1024, 1024, 1024, 1024, 1024, 1024, 1024};

float lastMsgTime;
float displayRefreshInterval = 20; // min time between screen draws

void setup() {
   String os=System.getProperty("os.name");
```

```
    println(os);
    initComms();
    fontA = createFont("Arial.normal", fontSize);
    textFont(fontA);
    textHeight = (int)textAscent();
    for (int i = 0; i <= maxNumberOfRows; i++)
      labelList.add(Integer.toString(i));
    adjustSize();
    drawGrid();
}

void adjustSize()
{
  topMargin = 3 * textHeight;
  bottomMargin = 0;
  if (displayWidth > 800) {
    windowWidth = 800;
    windowHeight = topMargin + bottomMargin + yPos(maxNumberOfRows);
    size(windowWidth, windowHeight);
  }
  else {
    windowWidth = displayWidth;
    windowHeight = displayHeight;
  }
  //leftMargin = getleftMarginLen() ;
  graphHeight = windowHeight - topMargin - bottomMargin;
  rectCenter = leftMargin + graphWidth / 2;
  rectLeft = leftMargin;
  rectRight = leftMargin + graphWidth;
}

void drawGrid() {
  fill(0);
  String Title = "Arduino Data" + commsPortString() ;

  int xPos = (int)( rectCenter - textWidth(Title)/2) ;
  text(Title, xPos, fontSize*2); // Title

  line(rectLeft, topMargin + textHeight,
       rectLeft, yPos(maxNumberOfRows) + 2);      // left vertical line

  line(rectRight, topMargin + textHeight, rectRight, yPos(maxNumberOfRows )+ 2);  // right line
  line(rectCenter, topMargin+textHeight, rectCenter, yPos(maxNumberOfRows) + 2); // center line

  for (int i=1; i <= maxNumberOfRows; i++) {
    fill(0);
    text(labelList.get(i), 2, yPos(i));  // row labels
    fill(150);
    String rangeCaption = "(" + rangeMin[i] + "~"  + rangeMax[i] + ")";
    text(rangeCaption, rectRight + textWidth("  "), yPos(i)); // range caption
  }
}

int yPos(int index) {
```

```
    return topMargin  + ((index) * textHeight * 2);
}

void drawBar(int rowIndex) {
  fill(204);
  if ( rangeMin[rowIndex] < 0) {
    if (values[rowIndex] < 0) {
      int width = int(map(values[rowIndex], 0, rangeMin[rowIndex], 0, graphWidth/2));
      rect(rectCenter-width, yPos(rowIndex)-fontSize, width, fontSize);
    }
    else {
      int width = int(map(values[rowIndex], 0, rangeMax[rowIndex], 0, graphWidth/2));
      rect(rectCenter, yPos(rowIndex)-fontSize, width, fontSize);
    }
  }
  else {
    int width=int(map(values[rowIndex], rangeMin[rowIndex], rangeMax[rowIndex], 0,graphWidth));
    rect(rectLeft, yPos(rowIndex)-fontSize, width, fontSize);    //draw the value
  }
  fill(0);
  text(values[rowIndex],
       rectRight + (int)textWidth(" (-1000~1000)    "), yPos(rowIndex)); // print the value
}

void processMessages() {
  while(true) {
    String message = commsGetMessage();
    if (message.length() > 0)
    {
      int row = 0;

      String [] data  = message.split(","); // Split the CSV message
      if ( data[0].equals("Data")) {  // check for data header
        row =  Integer.parseInt(data[1]);
        values[row] = Integer.parseInt(data[2]);
        checkRefresh();
      }
      else if ( data[0].equals("Label") ) { // check for label header
        row =  Integer.parseInt(data[1]);
        labelList.set(row, data[2]);
        if ( (int)textWidth(data[2]) > leftMargin) {
          leftMargin = (int)(textWidth(data[2]) + textWidth(" ") + 2) ;
          adjustSize();
        }
        checkRefresh();
      }
      else if ( data[0].equals("Range")) {  // check for Range header
        row =  Integer.parseInt(data[1]);
        rangeMin[row] = Integer.parseInt(data[2]);
        rangeMax[row] = Integer.parseInt(data[3]);
        checkRefresh();
      }
      else
        println(message) ;
```

```
    }
    else
      break; // finish processing when the message length is 0
    }
  }

  void checkRefresh()
  {
    if ( lastMsgTime < 1)
      lastMsgTime = millis(); // update the time if it was reset by the last display refresh
  }

void draw() {
  processMessages();
  if ( millis() - lastMsgTime > displayRefreshInterval)
  {
    background(255);
    drawGrid();
    for ( int i=1; i <= maxNumberOfRows; i++)
    {
      drawBar(i);
    }
    lastMsgTime = 0;
  }
}


/*******************************
   code for Serial port
*****************************/

import processing.serial.*;

Serial myPort;   // Create object from Serial class

void initComms(){
  String portName = Serial.list()[portIndex];
  println(Serial.list());
  println(" Connecting to -> " + portName) ;
  myPort = new Serial(this, portName, 9600);

}

String  commsPortString() {
  return " (" + Serial.list()[portIndex] + ")"  ;
}

String message;

String commsGetMessage() {

  if (myPort.available() > 0) {
    try {
      message = myPort.readStringUntil(10);
      if (message != null) {
```

```
       // print(message);
         return message;
        }
     }
   }
   catch (Exception e) {
      e.printStackTrace(); // Display whatever error we received
   }
 }
 return "";
}
```

This sketch talks to Arduino using the serial port and you need to ensure that the Processing sketch is using the same port that is connected to your robot. The port Arduino uses is displayed on in the Arduino IDE. You set the Processing port by changing the value of the variable portIndex. When starting the Processing sketch, you will see a list of the ports on your computer. portIndex is the position of the Arduino port in this list, but note that the index starts from 0, so the default value of 1 for portIndex is for the second port in the list.

A robot tethered via USB is not very convenient when you want to see what the robot is doing while moving. Adding a wireless serial device such as Bluetooth or XBee can be a big help when debugging or tuning your robot. If you are using a Leonardo, note that the TX/RX pins (digital 1 and 0) are accessed through Serial1 rather than Serial, so modify your code accordingly (you'll need to replace all instances of Serial with Serial1 in all the tabs of your sketch).

A standard board like the Uno uses the same Serial object as USB and although you don't need to modify the example code, you will need to disconnect the wireless device from the pins when uploading code. This is because the wireless device uses the same pins (digital 1 and 0) as USB.

Power Sources | D

Monitoring Battery Voltage

The battery voltage can be monitored using an Arduino analog input, but you can't directly connect the battery to an input pin because a fully charged battery can exceed the maximum voltage that the Arduino chip can tolerate.

Another factor to be aware of is that the default voltage reference for analog Read is the 5 volt output from the regulator on the Arduino board. This regulator requires more than 6 volts to produce a stable 5 volt output. When the voltage difference between the regulator input and output (referred to in the regulator datasheet as the *dropout voltage*) is less than a volt, the output voltage will drop below the required 5 volt level. Because this voltage is used as the default Arduino reference for analog conversion, the analog readings will no longer be accurate. In short, you shouldn't rely on the battery voltage as a reference to measure the battery voltage. So you need a reliable voltage reference that is not dependent on the output from the regulator.

The solution is to use an internal voltage reference that is built into the Arduino chip. This provides a 1.1 volt reference that is stable for any voltage that is sufficient to power the Arduino chip. Because the reference is 1.1 volts, the voltage being measured must not exceed this value, so a voltage divider to drop battery voltage down to an acceptable range is required (Figure D-1).

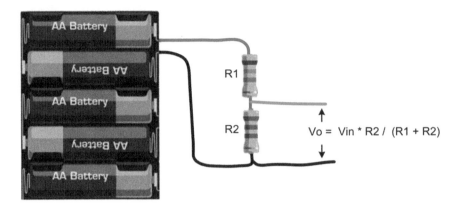

Figure D-1. *Resistors used as a voltage divider*

To support a wide range of battery choices (including 8.4 volt LiPo batteries), resistor values of 18k ohms for R1 and 2.2k ohms for R2 provide a voltage range of up to 10 volts.

Here is the voltage divider formula for these resistor values: R2 / R1 + R2. Substituting the chosen values results in:

`2200/(18000 + 2200)`

`= 0.109`

Therefore the voltage on the terminal will be the battery voltage times 0.109. For example, 10 volts at the battery will be dropped to just under the 1.1 volt range of the internal reference.

The resistors can be attached to the battery terminals as shown in Figure D-2, but a more permanent solution is to solder the resistors to the shield as shown in Figure D-3 and Figure D-4.

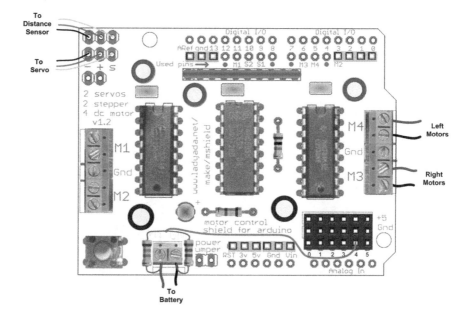

Figure D-2. *Resistors added to shield to monitor battery*

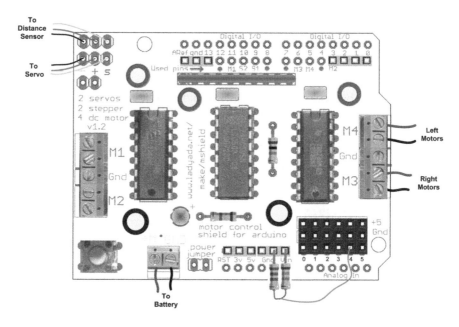

Figure D-3. *Voltage Divider Resistors soldered to Vin and Gnd pins*

Figure D-4. *Voltage Divider Resistors soldered to Vin and Gnd pins*

The code to read and interpret the voltage is in the Battery tab (Example D-1). This code reads the output of the voltage divider using analogRead and converts this into the battery voltage expressed in millivolts. This is compared to preset thresholds levels so an LED can be flashed to indicate low and critical battery levels. The code can also detect if the optional charger plug is connected to stop robot movement while being recharged.

Example D-1. **Battery tab code**

```
// code to monitor battery voltage

/*******************************************************************
 * LED starts flashing when volage drops below warning level
 * mark space ratio increses from 10% to 50% as voltage decreses from warning to critical
 * robot shuts down when battery below critical and led flashes SOS
 *
 * LED mark space ratio changes from 10% to 90% as voltage increases to full
 *******************************************************************/


// thresholds are the cell millivolts times number of cells
```

```
const int batteryFull    =  1500 * 5; // threshold for battery is low warning
const int batteryWarning =  1100 * 5; // threshold for battery is low warning
const int batteryCritical=  1000 * 5; // threshold to shut down robot

int batteryMonitorPin;      // analog pin to monitor
int chargerDetectPin =-1;   // pin goes open circuit when charger connected, default is no pin
int blinkPin;               // led pin to flash


void batteryBegin(int monitorPin, int ledPin)
{
  batteryMonitorPin = monitorPin;
  blinkPin = ledPin;
  pinMode(blinkPin, OUTPUT);
}

// version for charger detection
void  batteryBegin(int monitorPin, int ledPin, int chargerPin)
{
  batteryBegin(monitorPin, ledPin);
  chargerDetectPin = chargerPin;
  pinMode(chargerDetectPin, INPUT_PULLUP); // connect pull-up resistor
}

// indicates battery status using the given LED
void  batteryCheck()
{
  int mv = batteryMv(batteryMonitorPin); // get battery level in millivolts
  Serial.print("mv="); Serial.print(mv);
  if(chargerDetectPin >=0 &&  digitalRead(chargerDetectPin) == HIGH)
  {
    // here if charger detect is enabled and charger plugged in
    while( digitalRead(chargerDetectPin) == HIGH) // while charger is plugged in
    {
        moveStop();
        mv =  batteryMv(batteryMonitorPin); // get battery level in millivolts
        Serial.print(", charger detected, voltage=");
        Serial.println(mv); Serial.println(", percent=");
        int percent = map(mv, batteryCritical, batteryFull, 50, 100);
        percent = constrain(percent, 0, 100);
        Serial.println(percent);
        flash(percent, blinkPin);
    }
  }
  else
  {

    if(mv < batteryCritical)
    {
      Serial.println("Critical");
      // shut down the robot
      moveStop();
      while(1) {
        flashCritical(blinkPin);
```

```
            // check of the charger is plugged in
            if(chargerDetectPin >=0 && digitalRead(chargerDetectPin) == HIGH)
                return; // exit if charging

            delay(5000);
        }
    }
    else if (mv < batteryWarning)
    {
        int percent = map(mv, batteryCritical, batteryWarning, 10, 50);
        flash(percent, blinkPin);
    }
  }
  delay(1000);
  Serial.println();
}

// return the voltge on the given pin in millivolts
// see text for voltage divider resistor values
int  batteryMv(int pin )
{
#if defined(__AVR_ATmega32U4__) // is this a Leonardo board?
  const long INTERNAL_REFERENCE_MV = 2560; // leo reference is 2.56 volts
#else
  const long INTERNAL_REFERENCE_MV = 1100; // ATmega328 is 1.1 volts
#endif
  const float R1 = 18.0;  // voltge dividier resistors values, see text
  const float R2 = 2.2;
  const float DIVISOR = R2/(R1+R2);

  analogReference(INTERNAL);        // set reference to internal (1.1V)
  analogRead(pin);   // allow the ADC to settle
  delay(10);

  int value = 0;
  for(int i=0; i < 8; i++) {
    value = value + analogRead(pin);
  }
  value  = value / 8; // get the average of 8 readings
  int mv = map(value, 0,1023, 0, INTERNAL_REFERENCE_MV / DIVISOR );

  analogReference(DEFAULT); // set the reference back to default (Vcc)
  analogRead(pin); // just to let the ADC settle ready for next reading
  delay(10); // allow reference to stabalise

  return mv;
}

// flashes SOS in morse code
void flashCritical(int pin)
{
  for(int i=0; i< 3; i++)
    flash(20, pin);
  for(int i=0; i< 3; i++)
```

```
      flash(60, pin);
   for(int i=0; i< 3; i++)
      flash(20, pin);
}


// percent is the percent of on time time (duty cycle)
void flash(int percent, int pin)
{
   Serial.print(", flash percent="); Serial.println(percent);
   const int duration = 1000;
   // Blink the LED
   digitalWrite( pin, HIGH);
   int onTime = map(percent, 0, 100, 0, duration);
   delay(onTime);
   digitalWrite( pin, LOW);
   delay(duration - onTime);
}
```

There are two versions of the batteryBegin function. Use the one with three parameters if you have wired up the trickle charger circuit. The three parameters passed to the function are: the pin that the voltage divider is connected to, the LED pin, and the pin that detects the charger plug. Here is the function:

```
batteryBegin(alogBatteryPin, ledPin, chargerDetectPin)
```

If you have not wired the robot to use a charger, then call batteryBegin with two parameters: the pin that the voltage divider is connected to and the LED pin:

```
batteryBegin(alogBatteryPin, ledPin)
```

The checking is done in the batteryCheck function. This gets the battery level in millivolts by calling batteryMv and compares this to the warning and critical thresholds. The LED is flashed when the level drops below the warning level with a flash ratio (blink on time to off time) that changes as the voltage drops. If the voltage drops below the critical level, the robot movement is stopped, and the LED flashes a distress signal (SOS in morse code) every 5 seconds. When this happens, the batteries must be replaced or recharged before the robot will reactivate.

The myrobotBatteryMonitor example sketch (Example D-2) in the download shows how to use the battery monitor function.

Example D-2. Battery monitor example sketch
```
/*********************************************************************
myRobotBatteryMonitor.ino
```

```
sketch to demonstrate battery voltage monitoring
based on myRobotWander

Robot wanders using forward scanning for obstacle avoidance
LED blinks when battery runs low, robot goes to sleep when battery is critical.

Created by Michael Margolis 22 July 2012
*****************************************************************************/
#include "robotDefines.h"  // global defines

#include <AFMotor.h>  // adafruit motor shield library
#include "RobotMotor.h"    // 2wd or 4wd motor library

const int ledPin = 13;          // onboard LED
const int alogBatteryPin = 5;    // input on analog 5
const int chargerDetectedPin = 2; // digital pin 2

// Setup runs at startup and is used configure pins and init system variables
void setup()
{
  Serial.begin(9600);
  blinkNumber(8); // open port while flashing. Needed for Leonardo only

  lookBegin();
  moveBegin();
  //batteryBegin(alogBatteryPin, ledPin);
  batteryBegin(alogBatteryPin, ledPin, chargerDetectedPin);

  pinMode(ledPin, OUTPUT);
  Serial.println("Ready");
}

void loop()
{
//  roam();
  batteryCheck();
}

// function to indicate numbers by flashing the built-in LED
void blinkNumber( byte number) {
    pinMode(LED_PIN, OUTPUT); // enable the LED pin for output
    while(number--) {
      digitalWrite(LED_PIN, HIGH); delay(100);
      digitalWrite(LED_PIN, LOW);  delay(400);
    }
}
```

Trickle Charging

The build chapters in the beginning of the book described a simple trickle charger that you can use to recharge NiMH batteries. This section describes how to use the charger as well as some important points to ensure that you don't damage your batteries.

Trickle charging is a method of recharging NiMH batteries that provides a slow but steady charging current which should fully recharge 5 AA cells in around 14 to 16 hours. The charger has been designed for cells with a rated capacity of 2000 to 2500 mAh (milliampere hours). Cells with a higher rating can be used but they will require a longer charging period.

Do not try to charge non-rechargeable batteries.

The batteries start charging when a DC power supply is plugged into the charging socket and the power switch is turned on. The charging circuit is designed for use with a 12 volt supply with a 2.1mm plug (positive on the center connector). Cells with the suggested rating should handle the trickle charge current for long periods, however it is good practice to keep your charge session to 24 hours or less, particularly if your DC supply could be delivering a little more than the recommended 12 volts.

Programming Constructs

E

The code in this book takes advantage of a number of Arduino functions that are summarized in this appendix. See the online Arduino reference for each function if you want more detail.

Digital I/O

pinMode(pin, mode);
> Configures a digital pin to read (input) or write (output) a digital value; see *http://arduino.cc/en/Reference/PinMode*

digitalRead(pin);
> Reads a digital value (HIGH or LOW) on a pin set for input; see *http://arduino.cc/en/Reference/DigitalRead*

digitalWrite(pin, value);
> Writes the digital value (HIGH or LOW) to a pin set for output; see *http://arduino.cc/en/Reference/DigitalWrite*

pulseIn(pin, pulseType, timeout);
> Returns the pulse width in microseconds of a changing digital signal on the given pin. pulseType (either HIGH or LOW) determines if duration is for a high or low pulse. timout is an optional value indicating how long to wait for a pulse (the default is one second); see *http://arduino.cc/en/Reference/PulseIn*

Analog I/O

```
analogRead(pin);
```
Reads a value from the specified analog pin. The value ranges from 0 to 1023 for voltages that range from 0 to the reference voltage (5 volts by default, but can be changed by using analogReference; see *http://ardui no.cc/en/Reference/AnalogRead*

```
analogReference(type);
```
Configures the reference voltage used for analog input. This is used in the battery monitor code discussed in Appendix D, *Power Sources*; see *http://arduino.cc/en/Reference/AnalogReference*

Math functions

```
min(x,y);
```
Returns the smaller of two numbers; see *http://arduino.cc/en/Reference/Min*

```
max(x,y);
```
Returns the larger of two numbers; see *http://arduino.cc/en/Reference/Max*

```
constrain(x,lower,upper);
```
Constrains the value of x to be between the lower and upper range; see *http://arduino.cc/en/Reference/Constrain*

```
map(x,fromLow,fromHigh,destLow,destHigh);
```
Scales a value from one range to another range. The result will have the same proportion within the destination range as in the source range. The following code scales the analogRead value to a percentage of the full scale reading:

```
int val = analogRead(0);
int percent = map(val, 0,1023, 0,100)
```

The following code scales an analogRead value to its value in millivolts (refMv is the reference voltage expressed in millivolts):

```
int mV = map(val, 0,1023, 0, refMv);
```

See *http://arduino.cc/en/Reference/Map*

Other Functions and Constructs

```
switch / case statements
```
Controls program flow by testing if a number matches one of a number of alternative values. Here is a simplified example from the remote control sketch that uses switch to execute the appropriate function associated with each command:

```
void processCommand(int command)
{
  switch(command)
  {
    case MOVE_LEFT    :    moveLeft();       break;
    case MOVE_RIGHT   :    moveRight();      break;
    case MOVE_FORWARD :    moveForward();    break;
    case MOVE_BACK    :    moveBackward();   break;
    case PIVOT_CCW    :    moveRotate(-90);  break;
    case PIVOT_CW     :    moveRotate(90);   break;
    case HALT         :    moveStop();       break;
  }
}
```

The break statement is necessary to prevent execution falling through to the following case statement. See *http://arduino.cc/en/Reference/Switch Case*

array

An array is a collection of variables accessed using an index number. The first element of an Arduino array is accessed using an index of 0. An array can be initialized when it is declared by placing values in curly brackets. The following declares an array named motorSpeed with two elements that will store the speed for the left and right motors and initialize the speed values to 0:

```
const int NUMBER_OF_MOTORS = 2;
int  motorSpeed[NUMBER_OF_MOTORS]  = {0,0}; // motor speed stored here (0-100%)
```

see: *http://arduino.cc/en/Reference/Array*

#include "header.h"

This makes functions and variables declared in the specified file available to your sketch. See *http://arduino.cc/en/Reference/Include*

Arduino Pin and Timer Usage

F

The tables in this section show the pin and timer resources used by the projects in this book. You can use the same pin assignments for the Leonardo boards or the standard ATmega328 boards such as the Uno. However, there are subtle low level differences between these boards, so if you are adding capabilities that use additional pins or resources beyond those described in this book, then check the documentation on pin and resource usage for your board.

Handling Resource Conflicts

The Arduino chip has a rich collection of hardware resources, but you can run up against a conflict if a feature you are adding requires a hardware resource that some other feature is already using. A resource conflict occurs when a function reconfigures or requires exclusive access to some hardware capability. Running out of analog or digital pins is one kind of resource conflict, usually easy to spot.

More subtle is a conflict caused by a library that requires a resource such a hardware timer that is already used by some other function. For example, a motor shield uses PWM to control motor speed and each motor requires a timer component. Arduino tries to hide the underlying hardware (one of the things that makes it easy to use) but this can result in things going wrong when a resource conflict does occur. Sometimes the compiler will report a problem with an error message about a resource conflict. But sometimes the sketch will compile without an error message even though a resource conflict is preventing the code from functioning as expected.

For example, the infrared remote control library uses a timer to decode pulses in the background. If this is the same timer used by some other function, say the Arduino Servo library, one or both of these libraries will malfunction. The solution is to either reassign one of the libraries to use a different timer, or to find an alternative way to perform one of the functions without a timer. Both of these approaches will be discussed in this appendix.

Modifying a Library to Change Timer Allocation

Modifying a library is not a task for a beginner, but some libraries are designed to allow configuration. For example, the irRemote library used in Chapter 11, *Remote Control* has a file named irRemoteInt.h that can be edited to change the timer used by this library. Here are fragments of this file that determines the timer used by the library:

```
// Leonardo or Teensy 2.0
#elif defined(__AVR_ATmega32U4__)
  //#define IR_USE_TIMER1    // tx = pin 14
  // #define IR_USE_TIMER3    // tx = pin 9
  #define IR_USE_TIMER4_HS  // tx = pin 10
```

And further down the file:

```
// Arduino Duemilanove, Diecimila, LilyPad, Mini, Fio, etc
#else
  //#define IR_USE_TIMER1    // tx = pin 9
  #define IR_USE_TIMER2      // tx = pin 3
#endif
```

The first code fragment determines the timer to be used with a Leonardo board (the Arduino build process will use the code in this fragment if the chip is an ATmega32U4). The uncommented line contains: #define IR_USE_TIMER4_HS which results in the library using Timer 4. However, Timer 4 is also used to control one of the motors in the 4WD robot. If you have the 4WD robot and want to use the infrared remote control library, you need to find a free timer to use. You can't easily change the motor library because the pin for Timer 4 is hard wired to the motor controller chip. But you can change the remote timer by commenting out the line for Timer 4 and uncommenting a line that enables a free timer. The Leonardo has 5 timers but as shown in Table F-2, only Timer 1 is available. The code to disable Timer 4 and enable Timer 1 is as follows:

```
// Leonardo or Teensy 2.0
#elif defined(__AVR_ATmega32U4__)
  #define IR_USE_TIMER1     // tx = pin 14
  // #define IR_USE_TIMER3     // tx = pin 9
  // #define IR_USE_TIMER4_HS  // tx = pin 10
```

Using your text editor to make and save that change in irRemoteInt.h will eliminate the conflict by using Timer 1 instead of Timer 4.

If your 4WD robot uses an Arduino Uno, then the change is to remove the // comment characters before the IR_USE_TIMER1 line and add the comment characters before IR_USE_TIMER2

```
// Arduino Duemilanove, Diecimila, LilyPad, Mini, Fio, etc
#else
  #define IR_USE_TIMER1   // tx = pin 9
  //#define IR_USE_TIMER2    // tx = pin 3
#endif
```

Writing Code That Avoids the Use of a Timer

Sometimes there is a conflict but no alternative resource to use. An example of this is if your infrared remote library is using Timer 1 (see previous section) and you also want to use the Servo library, which also uses Timer 1. If you have the 2WD robot with a Uno, then you could use Timer 2 for the remote library so the Servo library can remain on timer 1. There are no free timers available if you have a 4WD robot or the 2WD robot with a Leonardo board, but you can solve this conflict by adding some code that controls the servo without using a timer. See "Adding Scanning" (page 178) for an example of how this can be done.

Pin and Timer Tables

The best way to handle hardware conflicts is to plan in advance by familiarizing yourself with the resources currently in use and the resources needed by the function you are adding. The tables in this appendix show the chip pins and timers used by the projects in this book. Although you will have some pins free after connecting up all the projects presented in this book, there may not be enough pins to connect all the optional sensors mentioned in Chapter 8, *Tutorial: Introduction to Sensors* along with some of the suggestions in the appendices. Use Table F-1 and Table F-2 to keep track of your pin and timer allocations.

Table F-1. Pin Usage

Pin	Usage	Comment
Digital 0		Serial Receive
Digital 1		Serial Transmit
Digital 2	Unused	Leonardo can use this for I2C
Digital 3	Motor 2 PWM	Timer 2b on Uno, Timer 0b on Leo (Leo uses this for I2C)
Digital 4	Motor control	
Digital 5	Motor 4 PWM	Timer 0b on Uno, Timer 3a on Leo
Digital 6	Motor 3 PWM	Timer 0a on Uno, Timer 4d on Leo
Digital 7	Motor control	
Digital 8	Motor control	
Digital 9	Scan Servo	used in Chapter 10, *Autonomous Movement*
Digital 10	Distance Sensor	used in Chapter 10, *Autonomous Movement*

Pin	Usage	Comment
Digital 11	Motor 1 PWM	Timer 2a on Uno, Timer 0a or 1c on Leo
Digital 12	Motor control	
Digital 13	On-board LED	This can be used as a digital pin if LED not needed
Analog 0	Left Reflectance Sensor	
Analog 1	Right Reflectance Sensor	
Analog 2	Center Reflectance Sensor	
Analog 3	IR Remote Decoder	used in Chapter 11, *Remote Control*
Analog 4	Optional Battery Monitor	Uno can use this for I2C
Analog 5	Optional sound or proximity sensor	Uno can use this for I2C

Table F-2. Timer Usage

Timer	Uno 2WD	Uno 4WD	Leo 2WD	Leo 4WD
Timer 0		PWM for motors 3 & 4	PWM for motor 1 & 2	PWM for motor 1 & 2
Timer1	IR Remote	IR Remote	IR Remote	IR Remote
Timer2	PWM for motors 1 &2	PWM for motors 1 &2	Not available	Not available
Timer3	Not available	Not available		PWM for motor 4
Timer4	Not available	Not available		PWM for motor 3

CPSIA information can be obtained at www.ICGtesting.com
Printed in the USA
BVOW101211161212

308313BV00001B/1/P

9 781449 344375